PSYCHOLOGY
RESEARCH METHODS

KEY SCIENCE SKILLS WORKBOOK

KRISTY KENDALL

THIRD EDITION

Psychology Research Methods: Key Science Skills Workbook
3rd Edition
Kristy Kendall
ISBN 9780170465038

Publisher: Eleanor Gregory
Project editors: Felicity Clissold, Alex Chambers, Alana Faigen
Editor: Kelly Robinson
Series text design: Ruth Comey (Flint Design)
Series cover design: Cengage Creative Studio (original design by Fiona Lee, Studio Pounce)
Series designer: Linda Davidson (Cengage Creative Studio)
Permissions researcher: Liz McShane
Production controller: Karen Young
Typeset by: Lumina Datamatics

Any URLs contained in this publication were checked for currency during the production process. Note, however, that the publisher cannot vouch for the ongoing currency of URLs.

For product information and technology assistance,
in Australia call **1300 790 853**;
in New Zealand call **0800 449 725**

For permission to use material from this text or product, please email **aust.permissions@cengage.com**

ISBN 978 0 17 046503 8

Cengage Learning Australia
Level 7, 80 Dorcas Street
South Melbourne, Victoria Australia 3205

Cengage Learning New Zealand
Unit 4B Rosedale Office Park
331 Rosedale Road, Albany, North Shore 0632, NZ

For learning solutions, visit **cengage.com.au**

Printed in China by 1010 Printing International Limited.
2 3 4 5 6 7 26 25 24 23

CONTENTS

ISBN 9780170465038

INTRODUCTION

One of the fundamental skills in Psychology is the ability to understand and apply research methodologies. This is not a straightforward task, because the material uses subject-specific language, and the concepts are interrelated, making it difficult to know where to begin.

The development of key science skills is a core component of the study of Psychology, not just for use in secondary education, but also in tertiary education and beyond. The ability of researchers and practitioners to understand how to conduct, critique and present robust research is what allows the discipline of Psychology to progress.

This workbook allows students to learn at their own pace. It may be used as a student-centred learning tool, with all the necessary content from the course covered in full detail. It can also be used as a teacher-driven resource, providing support in covering the concepts through class discussion or homework tasks. This workbook is unique because it allows students to learn one skill at a time and to apply their understanding of each skill as they progress through the course content.

This workbook enables students to develop knowledge, to understand the practical applications of the content and to then apply this understanding. Above all else, it is a workbook in which all aspects of this important topic are drawn together in a cohesive manner, and a resource that enables students to develop their key science skills to a progressively higher level with increased proficiency.

AUTHOR ACKNOWLEDGEMENTS

The author would like to thank her amazing school, Toorak College, and all of the students who inspire her to continually think about ways to help them to learn and engage with the material in the subject of Psychology. She would also like to thank her fellow educators and the teachers everywhere who continue to innovate and adapt their practice. Finally, she would like to thank her fabulous family – Lawrence, Hudson and Chase – who have always supported her dreams.

PUBLISHER ACKNOWLEDGEMENTS

The publisher would like to thank Amelia Brear and Chalsea Chappel, who provided insightful reviews and feedback on the manuscript.

Warning – First Nations Australians are advised that this book and associated learning materials may contain images, videos or voices of deceased persons.

1 DEVELOPING RESEARCH

1.1 The significance of research

Research is all around us – whether determining the safety of a new vaccination, exploring the best way to change people's opinions on global warming or simply determining the most effective way to influence a buyer when marketing a new product. When it comes to research in Psychology, researchers are interested in the human mind and its associated thoughts, feelings and behaviours.

When embarking on research you must first determine the types of questions you want to answer; for example, how do you convince people to become vaccinated? Why are mental health problems prevalent during adolescence? Can listening to content while you sleep really help you learn? From these questions, a research **aim** will emerge. An aim is a statement that explains what you intend to investigate; for example, you may aim to investigate the effects of the media on body image.

There are many ways to conduct research, and it is important to undertake the scientific **methodology** that will enable you to achieve your desired outcomes. Scientific methodologies are ways in which to systematically study or explore an area of interest. Some examples follow.

- **Case study**: This is an in-depth or detailed study of a particular person, activity, behaviour or event. Generally, case studies are linked to real-world events or scenarios.

- **Fieldwork**: This is an investigation of an issue or line of inquiry in its natural environment. Fieldwork can involve **observational studies** of people, or an investigation of people's opinions through questionnaires undertaken outside of a controlled environment.

- **Literature review**: This is the collation and analysis of the findings of others, known as **secondary data**. When conducting a literature review, the researcher brings together a collection of research and ideas, and draws comparisons and conclusions from them.

- **Controlled experiment**: This is a study that occurs under controlled conditions and investigates a cause-and-effect relationship between two or more **variables**. It also tests a **hypothesis**.

There are many other ways to conduct scientific investigations through modelling, **simulation** or the development of products and processes. As well as choosing appropriate methodology, we must also ensure that the research conducted is robust, and is aligned with the rules that govern research so it can be published and used as a vehicle for change. Psychological research must be conducted in accordance with ethical principles. The term **ethics** refers to the moral principles and codes of behaviour that apply to all psychologists in their practice and in their research. These ethical principles ensure the health and safety of anyone involved in research and help to ensure that the research is valuable.

Shutterstock.com/Kzenon

Nudges were used to encourage social distancing.

Nudges: four reasons to doubt popular technique to shape people's behaviour

10 January 2022   Author: Magda Osman

Throughout the pandemic, many governments have had to rely on people doing the right thing to reduce the spread of the coronavirus – ranging from social distancing to handwashing. Many enlisted the help of psychologists for advice on how to "nudge" the public to do what was deemed appropriate.

'Nudges: four reasons to doubt popular technique to shape people's behaviour' by Magda Osman, *The Conversation*, January 10, 2022. Adapted with permission from Magda Osman.

Figure 1.1 'Nudging' – a group of techniques used to promote a certain kind of behaviour through soft interventions and reminders rather than through mandates – was used during the COVID-19 pandemic. However, 'nudging' can present ethical concerns because it can target emotional, social and cognitive processes.

1 Brainstorm some questions from the world of Psychology that you are interested in finding the answers to. Think about the types of questions that psychologists would be interested to learn more about in relation to stress, sleep, mental illness, antisocial behaviour, behaviour change and so on.

2 When formulating an aim, we want to explore the effect of something on the subsequent outcome. Write a possible aim for this scenario: *A psychologist is interested in exploring why males are more aggressive than females.*

3 There are many different forms of scientific methodologies. For the scenarios below, identify an appropriate methodology that is aligned with the research aim.

a A researcher wants to explore existing research on the opinions of people in different countries about gender equality in the workplace.

b A researcher wants to explore whether drinking energy drinks leads to concentration problems in children.

c A researcher wants to explore whether students in single-gendered schools play differently during break times than students in co-educational schools.

1.2 Steps in research

Researchers, or experimenters, use a series of systematic steps to plan, conduct, interpret and report on the research they undertake. They do this regardless of the methodology they choose. Seven steps are commonly used in conducting psychological research. Read the following steps to gain a holistic understanding of how the research process progresses.

Step 1: Identify the research problem

The first step in psychological research is to identify an area that needs investigating. When a psychological researcher identifies a topic of interest, they must conduct a literature search. This involves finding and reading relevant research articles, results and journals on the topic. Once they have completed the literature search, the researcher must develop a testable research question or problem. For example, a researcher may be interested in investigating the causes of, or influences on, eating disorders in adolescent females. The researcher would do a literature search on the broad topic, then identify a research question, such as: *Is the prevalence of eating disorders in adolescent girls affected by whether they have a relative who either currently has an eating disorder, or has had an eating disorder in the past?*

Step 2: Formulate a hypothesis

The next step in psychological research is to formulate a hypothesis. A hypothesis is a testable prediction about the relationship between two or more variables (events or characteristics). In other words, a hypothesis is an educated guess about the results of the experiment; for example, *It is hypothesised that adolescent girls who have, or have had, a relative with an eating disorder are more likely to have an eating disorder than those who do not have a relative with an eating disorder.*

A hypothesis is based on the information gathered in the literature search; it is not just a guess made without any prior knowledge or investigation. The hypothesis is developed before the research is conducted and is written as a very specific statement that includes the **population** from which the **sample** is drawn, the variables manipulated, and the anticipated outcome.

Step 3: Design the method

The third step is to design the **method** that will be used to undertake the research. The research method determines how the researcher will test the hypothesis. When designing the method, the researcher must determine who will participate in the study and how to sample them, how many **participants** are required for the study, what the participants will do and under what conditions, and what will be measured. This is also the stage in which researchers try to minimise any unwanted variables that may affect the research when it is being conducted.

Step 4: Collect the data

The fourth step is to collect the data. Data-collection procedures include controlled experiments, naturalistic observations, surveys, interviews and case studies. To measure participants' performance, opinions or results, a researcher can use a variety of data-collection techniques, including questionnaires, direct observation, psychological tests or examination of files and documents. The data collected can take many different forms.

Step 5: Analyse the data

Data analysis involves organising, summarising and representing the raw data in a coherent and logical manner. **Raw data** are the actual data collected from a study. The results of a research investigation will often comprise large amounts of raw data, and the researcher must make sense of these through statistical procedures and summaries. **Descriptive statistics** are used to describe, summarise and organise data. Examples of descriptive statistics include graphs, percentages, tables, measures of central tendency and frequency distributions. These will be discussed in Chapter 3.

Step 6: Interpret the results

Once the data have been summarised and organised, they must be interpreted. Interpreting results involves forming **conclusions** about what the data show. **Inferential statistics** are statistics that allow you to make inferences and conclusions about data and are often used to interpret results. Such statistics allow us to explain the significance of data in light of the variables manipulated. At this stage, a conclusion may be drawn, which is a decision or judgement about the meaningfulness of the results of a study.

Step 7: Report the findings

Psychological research is conducted so that the results can be shared with other psychologists and researchers. Therefore, the final step in psychological research is to report the data to others. After conducting a study, the researcher usually prepares a report that is either submitted to a journal, incorporated into a thesis or presented to other psychologists at a conference. Once a research report is published, other researchers may use it in their own literature searches and in further investigations. Publication also enables the general public to benefit from research findings.

ISBN 9780170465038

1 Fill in the flow chart to identify the seven steps in psychological research. Next to each step, give an example of what may be done or found in each step. The research problem has been identified for you.

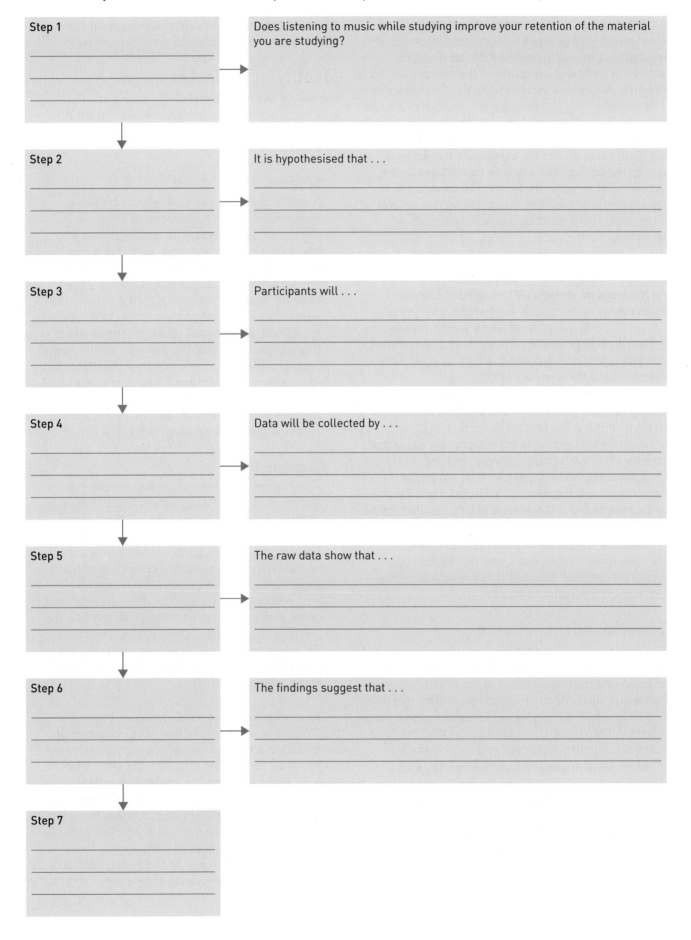

Step 1

_____

_____

_____

Does listening to music while studying improve your retention of the material you are studying?

Step 2

_____

_____

_____

It is hypothesised that . . .

_____

_____

_____

Step 3

_____

_____

_____

Participants will . . .

_____

_____

_____

Step 4

_____

_____

_____

Data will be collected by . . .

_____

_____

_____

Step 5

_____

_____

_____

The raw data show that . . .

_____

_____

_____

Step 6

_____

_____

_____

The findings suggest that . . .

_____

_____

_____

Step 7

_____

_____

_____

1.3 Independent and dependent variables

A controlled experiment investigates a cause-and-effect relationship between two or more variables; that is, whether a change in one thing has an impact on another. A variable is any condition that can change. For example, if research involved testing the effectiveness of a new memory drug on participants' retention of a list of nonsense words, two variables would be involved when conducting this experiment: whether or not participants are given the drug, and the retention of the nonsense words. The first variable would be manipulated by the experimenter: whether the participants are exposed to the experimental condition or not. The second variable is dependent on the participants; it is their results or responses.

The independent variable

An **independent variable (IV)** is a condition that an experimenter systematically manipulates, changes or varies in order to gauge its effect on another variable. If there are two groups of participants in an experiment, one group would be exposed to the IV, or experimental condition, and the other would not.

Consider an experiment in which one group of people is sleep deprived for 24 hours and another group is allowed to sleep normally. The IV is the factor that is different between the two groups; in this scenario, the amount of time for which each group was sleep deprived. To explain this completely, we must provide detail about how the variable will be manipulated or measured. Therefore, the IV could be stated as the amount of time for which each group was sleep deprived (0 hours or 24 hours).

Once the difference between groups is identified, the researcher then explores how the results of each group are different. The results in this case would be known as the dependent variable.

The dependent variable

The **dependent variable (DV)** is that which is measured in an experiment. Dependent variables measure the effect(s) that the manipulation of the IV has had on behaviour. Such effects are often revealed by measures of performance, such as test scores or number of goals scored. A DV 'depends' on another variable, the IV, to determine whether it changes and by how much.

In the previous sleep scenario, where one group was sleep deprived for 24 hours and the other group was allowed to sleep normally, what could experimenters have been measuring? They could have used various tests the next day to measure participants' concentration levels, their memory or even their mood – all of these are possible DVs.

The DV should be explained in terms of how it will be measured. For example, a DV for the sleep scenario may

have been percentage of correct scores on a memory test, or average self-reported mood on a 10-point scale.

Identifying variables

Put simply, the IV is identified by looking at what is different between the conditions, and the DV is identified by looking at what is being recorded or measured. Let's look at this in action.

Scenario A

A researcher examining the effects of a new treatment drug for schizophrenia divides a sample of participants with schizophrenia into two groups. He ensures that participants in one group take the new drug every day for a week and that participants in the other group have no treatment at all. He then measures the number of hallucinations the participants in each group reportedly experience over the week. In this scenario, the IV is the presence of the new drug and the DV is the number of hallucinations reportedly experienced in a one-week period. This experiment allows the researcher to see whether the presence of the new drug (IV) affects the experience of hallucinations (DV).

Scenario B

A researcher conducts a study into the effects of alcohol on driving. Participants in Group 1 drink five alcoholic beverages in one hour and participants in Group 2 drink five glasses of water in one hour. In the following hour, participants must navigate an obstacle course on a driving simulator, in which they must avoid hitting the traffic cones. In this experiment, the results will be measured by the number of traffic cones each participant hits. Therefore, the IV is the presence of alcohol and the DV is the accuracy on the driving simulator. This experiment allows the researcher to see whether the presence of alcohol (IV) affects driving ability (DV).

To ensure we can see the impact of the IV on the DV, we need to ensure that other variables are controlled and the two conditions are as similar as they can be; that is, the only variable should be the variable being examined. A **controlled variable** is any variable that is constant in research conditions. For example, in our experiment investigating the effect of alcohol on driving performance, controlled variables would include age/gender of participants, food consumed in the time prior to the experiment, and the type and timing of the driving simulation.

ISBN 9780170465038

1 Identify the independent and dependent variables in each experiment.

   a Professor Tallent is examining the effect of hunger on motivation in rats trying to run through a maze. She uses two groups of rats, and times how long it takes members of each group to successfully complete the maze to find a piece of food. Professor Tallent feeds the rats in one group before they enter the maze but does not feed the rats in the other group until they have completed the maze.

   Independent variable: _____

   Dependent variable: _____

   b Emma has always been told that she makes excellent muffins, but she wants to find out which of two flavours she makes is the best. For two months (January and February), Emma makes muffins and distributes them to her friends and family. She makes the same number of muffins each month and distributes them to the same people, but for the month of January she makes only blueberry muffins and for the month of February she makes only chocolate muffins. Emma counts the number of compliments she receives on her muffins over the two months.

   Independent variable: _____

   Dependent variable: _____

   c Luli was in charge of a large corporation that taught people foreign languages for travel purposes. She had equal numbers of students in two 6-week programs learning the same language: one where students learned by hearing and repeating the foreign words and another where students learned by reading and writing the foreign words. Luli decided to test which method was more effective by giving the students of each program a test at the end of the 6 weeks.

   Independent variable: _____

   Dependent variable: _____

2 Parents often tell children old wives' tales and stories to discourage them from displaying undesirable behaviours. Design an experiment to test the following well-known statements.

   a Sitting too close to a screen is detrimental to your vision.

   Experiment design: _____

_____

_____

_____

_____

_____

   Independent variable: _____

   Dependent variable: _____

   b Eating bread crusts makes your hair curly.

   Experiment design: _____

_____

_____

_____

_____

_____

   Independent variable: _____

   Dependent variable: _____

1.4 Formulating a hypothesis

Once a researcher has identified the variables at play in investigating a research question, they need to make an educated guess about the relationship between these two variables. A hypothesis is a statement, or testable prediction, about the likely outcome of an experiment.

A hypothesis makes a prediction about the direction of interaction between the IV and the DV (will it increase, decrease or stay the same?) and also includes the population from which the sample is drawn. For example, you might predict that sleep deprivation will lead to decreased memory ability in adolescent males. You may include detail on how much sleep or what types of memory ability are being referred to, or how memory ability might be measured.

It is important that any hypothesis you write includes:

1 a testable prediction about the direction of interaction between variables (i.e. higher, lower, increased, decreased etc.)
2 the population from which the sample is drawn
3 both conditions of the independent variable (that which is manipulated)
4 the dependent variable (that which is measured).

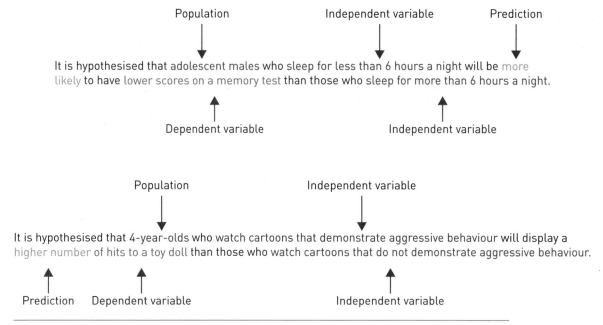

Figure 1.2 Two well-written hypotheses

ISBN 9780170465038

For each of the following scenarios, identify the independent variable and the dependent variable, and formulate a hypothesis for each.

1 Max is conducting research to investigate whether sleep quality is improved by meditating. He investigates a group of 40-year-old males, where half of the group practises meditation for 10 minutes before going to bed every night for one week, and the other half does not. Upon waking each morning for that week, the participants in both groups are to rate their quality of sleep on a scale from 1 to 10. An average of each participant's seven ratings is taken.

Independent variable: _____

Dependent variable: _____

Hypothesis: _____

_____

_____

_____

_____

2 When a Year 12 Psychology teacher taught her lessons in the first half of the year, she wrote notes on the board for her class to copy. She was disappointed in the results obtained by her class in the mid-year exam and wanted to investigate whether presenting her notes in a different way would impact on her students' results in the end-of-year exam. In the second half of the year, rather than write her notes on the board, she presented her notes in PowerPoint, and added lots of visual cues.

Independent variable: _____

Dependent variable: _____

Hypothesis: _____

_____

_____

_____

_____

3 Dr Shing is interested in assessing whether Victorians' memory abilities decrease with age. She places an advertisement in the newspaper calling for 20- to 30-year-olds and 50- to 60-year-olds to take part in her study. She then exposes participants in both age groups to a series of memory-related tests and compares the performances of the two groups using average scores obtained from the tests.

Independent variable: _____

Dependent variable: _____

Hypothesis: _____

_____

_____

_____

_____

1.5 Population and sample

Experimentation is the most widely used research method for learning about human behaviour. The people used in an experiment are called participants. The participants in a study are a selection, or sample, of people chosen from a particular population of research interest.

Population

A population is the entire group of people belonging to a particular category (for example, all university students or all AFL footballers). In experimental terms, it is the larger group of research interest from which a sample is to be drawn. In this context, the term *population* does not refer to the number of people living in a particular area (such as the population of Victoria), but rather a group of people with similar characteristics that are of interest to a researcher. For example, if you were to conduct a study investigating Victorian Year 12 students' favourite subjects, the population would be all students who are enrolled in Year 12 in schools in Victoria.

Sample

The group of participants in a research study are collectively called a sample. A sample is a group of participants selected from, and representative of, a population of research interest. A sample must represent the population from which it is drawn in order for inferences to be made about that population. The process of choosing participants from the population for use in a study is called *participant selection*, or *sampling*.

A sample is a subsection of a population, and therefore it is a smaller group than the population itself. Ideally, psychologists would like to examine all members of the population to obtain the most accurate results;

however, this is often impossible as there may be thousands, if not millions, of people in a particular population. Therefore, testing every member of a population would take too long, and be too expensive; this is why samples are taken from the population. Samples are most commonly obtained randomly; for example, by drawing names at random from a database. You will learn more about different sampling techniques in Chapter 2.

Participant allocation

Once participants have been selected for an experiment, they must be allocated to each of the groups within the experiment. As with sampling, the allocation (or assignment) of participants must be done in a systematic and carefully planned manner to ensure that participants' individual characteristics are evenly distributed among the groups. The best way to ensure that this happens is through **random allocation**. Random allocation is a technique that ensures that every member of the sample has an equal chance of being assigned to either of the groups used in the experiment. This may be done by placing the names of all members of the sample in a container and then drawing them out one by one, allocating half to each group.

An experiment usually has two types of groups: the **control group** and the **experimental group**. The experimental group is the group (or groups) exposed to the experimental condition(s); that is, where the variable being manipulated – the IV – is present. For example, in a test of the effects of a new drug, the experimental group is given that drug. Sometimes it is necessary to have different levels of the IV under investigation. There may be different strengths of drugs trialled; in this case you may see multiple experimental groups.

The control group is the group that is exposed to the control condition; that is, where the variable under investigation is absent. For example, in a test of the effects of a new drug, the control group is often given no treatment, or is given a fake pill or fake treatment rather than the new drug. A fake pill or treatment is known as a placebo, and will be discussed in more depth in Chapter 2.

The control group provides a basis of comparison, so that the performance of the experimental group can be compared with a base level. The control group allows the researcher to determine whether the drug (IV) has had an effect on the experimental group's behaviour (DV). Without the control group, an experimenter would have no idea whether the IV has had an effect on the DV or whether the change would have occurred naturally, or is due to other factors altogether.

iStock.com/Rawpixel

Figure 1.3 A sample is a subset of participants taken from the population of interest.

ISBN 9780170465038

1 Complete the flow chart by defining the terms.

Population:
_____
_____
_____

↓

Sample:
_____
_____
_____

↓

Random allocation:
_____
_____
_____

Control group:
_____
_____
_____

Experimental group:
_____
_____
_____

2 Why is it necessary to have a control group?

_____
_____
_____

3 If we were conducting an experiment on whether listening to music increases the enjoyment level of eating burgers at a takeaway franchise, and we wanted three experimental groups and one control group, what could each of the groups do?

Control group: _____
_____

Experimental group 1: _____
_____

Experimental group 2: _____
_____

Experimental group 3: _____
_____

2 MINIMISING EXTRANEOUS VARIABLES

2.1 Extraneous variables

Experiments aim to investigate whether the independent variable (IV) has an impact on the dependent variable (DV), but determining this can often be difficult because other variables may impact on the results. An **extraneous variable (EV)** is any variable *other than* the IV that can cause a change in the results and therefore has an unwanted effect on an experiment. EVs often interfere with the **causal** link between IVs and DVs; it can be difficult to determine whether the IV was responsible for a change in the DV, or whether the EV was responsible. Because of this, it is necessary for researchers to try to prevent EVs occurring.

One of the main intentions of an experiment is to ensure that, aside from the IV, the control and experimental groups are identical in their characteristics, conditions and procedures. There are many different types of EVs that can prevent these two groups from being comparable. EVs may include differences among research participants between the two groups. These are known as individual *participant differences* (for example, differences in terms of memory, motivation, mood, personality, expectations and ability). There may be differences in how the experimenter treats the participants, known as **experimenter effects**.

It is an experimenter's job to ensure that EVs are minimised. There are many ways to do this, including considering how a sample is obtained and how participants are allocated to control and experimental groups, considering the experimental design that is to be used, and considering how to minimise the effect of participant and experimenter expectations. While a researcher must try to minimise the effect of EVs, sometimes EVs may be present without the researcher knowing, and they may not be identified until after the experiment is complete, if at all.

It is important to remain critical when reviewing research and to look for potential problems with the research design and delivery. There may be differences in the way that a test is delivered or administered to participants; this is known as **non-standardised procedures**. It is therefore important to standardise or control each test and its procedures, which means ensuring that the test and all test conditions are the same each time the test is administered. For example, if you are testing two groups on basketball-shooting ability, you must make sure that both groups shoot in the same weather conditions, otherwise the wind or another weather factor – rather than the intended IV – may affect shooting accuracy.

Other extraneous variables that can influence results are things such as **artificiality**; that is, the unnatural environment in which an experiment is conducted. This could be avoided by designing an experiment to eradicate this factor; for example, by doing an observational study through fieldwork.

There are many types and categories of extraneous variables, more of which will be discussed throughout this chapter.

Confounding variables

If EVs are not controlled for, they can in some cases have so much of an impact that they have a confounding effect on the interpretation of results. This would mean that an EV, rather than the IV, causes a change in the DV. When an EV has a confounding effect on the results, it is known as a **confounding variable**. A confounding variable is an uncontrolled variable – i.e. not the specified IV – that has caused a change in the DV, and its effects on the results may be confused with the effects of the IV. That is, we know that the IV alone has not caused the change in the DV and instead there is an alternative explanation for the results.

For example, when conducting a study to investigate whether boys or girls are better at maths, a researcher may take a sample of girls aged 16–17 and a sample of boys aged 14–15. In this case the age of the students is a confounding variable. The differences between the participants in the two groups should be consistent in as many ways as possible, so that the effect of the IV, in this case gender, can be examined. As the two groups are also different in age, age itself may cause one group to perform better, instead of the intended IV (gender).

Figure 2.1 When trying to investigate the effect of gender on maths performance it is important to ensure participants are similar in other characteristics, such as age.

ISBN 9780170465038

1 Identify two possible extraneous variables in each of the scenarios.

   a Ravi is trying to see whether a new brand of tennis racquet improves the accuracy of a tennis serve. In her test, she uses two groups of participants on an outdoor court on Friday and Saturday nights; one group each night. Participants in the Friday-night group serve with their own racquets and participants in the Saturday-night group serve with the new racquet.

   _____

   _____

   _____

   b Dr Bennett has been commissioned to look at the possibility of lowering the legal blood alcohol level for drivers of motor vehicles. He samples one group of university students on a Monday and gives them four drinks in one hour. He samples another group on a Tuesday and gives them three drinks in one hour. He then has each group take a test on a driving simulator and he counts the number of errors they make.

   _____

   _____

   _____

   c Ms Luk-Tung is relatively new to the teaching profession and is interested in learning whether she is better suited to teaching boys or girls. She decides to give her two Biology classes (which contain a mix of girls and boys) a written survey in which she asks students to comment on her teaching abilities. She also asks that students write their names on the top of their surveys, so the surveys will not be anonymous.

   _____

   _____

   _____

2 Apply what you have learnt about extraneous and confounding variables to show the differences and similarities between them.

Extraneous variable   **Confounding variable**

2.2 Types of sampling

Psychologists use a number of different methods to sample participants for a research study. The sampling step is very important, because the main aim at the conclusion of the study is to apply the results from the sample to the larger population from which it was drawn. With this in mind, it is important that the sample is representative of the population. Sampling appropriately can also help to minimise the impact of EVs. There are many different ways to sample, including **convenience sampling**, **random sampling** and **stratified sampling**.

Convenience sampling

Convenience sampling, as its name suggests, is a quick and easy way of selecting participants. It involves selecting participants based on the researcher's accessibility to them, or the participants' availability; for example, sampling only one class in a school, or going to the local supermarket and surveying the people found there.

The main advantage of this sampling type is obvious: it is convenient. It also does not require forward planning and is quick to administer, which in turn minimises costs. However, this sampling technique can be highly biased. Participants may not necessarily be representative of the population because they are likely to share a particular quality, or be predisposed to act in a particular way. For example, if you wanted to study personalities of school students but you sampled only a drama class at a school, those particular students may be more creative or outgoing than the rest of the school's population, so the sample would not accurately represent the underlying population.

Alamy Stock Photo/Jeffrey Isaac Greenberg 18+

Figure 2.2 Convenience sampling, such as surveying people who attend a shopping centre, may not represent the wider population.

Random sampling

Random sampling employs a carefully planned and systematic method of selecting participants for a study. Random sampling ensures that every member of a population has an equal chance of being selected for the sample being used in the study. Two common

methods of random sampling are pulling names out of a container (such as a hat) or allocating a number to each person in the population and then using a random number generator to select the sample.

The main advantages of random sampling are that it is quick and it is inexpensive, as the sampling procedures are not difficult to set up. Additionally, and probably most importantly, random sampling is not biased – every member of the population has an equal chance of being selected to be part of the sample. In this way, random sampling is much more likely to be representative of the population than convenience sampling.

There is, however, the chance that the sample obtained through random sampling may not be representative of the population of research interest. Although the sample is chosen at random, by chance it may turn out that the sample is biased in one direction; for example, it may be entirely made up of men, or only of people aged more than 45 years old.

Stratified sampling

Stratified sampling involves breaking the population into 'strata' (singular: stratum), or groups, based on characteristics they share. For example, you could divide a secondary school population into year levels; these would be the strata. Once the population is divided, you then select participants randomly from each strata in the same proportions that they appear in the population.

Samples chosen using this technique should be representative of the population, because there should be equal quantities of members from each stratum in the ratio in which it appears in the population. For example, if there are more boys than girls in a school, a stratified sample of this population would also include more boys than girls. However, this type of sampling can be time-consuming to undertake, as we need information about the population's characteristics before sampling can begin.

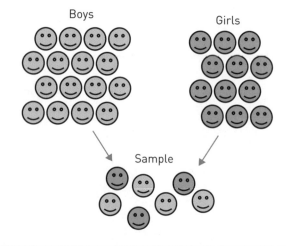

Figure 2.3 Stratify a sample by selecting members of each gender for the sample in the ratios in which they appear in the population.

ISBN 9780170465038

1 You want to determine the opinions of people in Victoria on sustainable practices in the home. Outline three ways in which you could obtain a convenience sample.

_____

_____

_____

_____

2 Your school has just decided that it will not hold a Year 12 formal this year because students were badly behaved at last year's formal. You would like to make a presentation to the principal showing the percentage of Year 12 students in the school who are against this decision. In order to gather information, you decide to have a random sample of 20 students in the school complete a survey. Outline three ways in which you could gather a random sample.

_____

_____

_____

_____

3 The employees of a company are having trouble agreeing on a venue for their Christmas party. The social coordinator decides to hold a vote on preferred venues, but the company is too large to ask all employees to vote. She decides to form a stratified sample, and only members of the sample will cast a vote. Explain how she could execute this to ensure the choice of venue is fair.

_____

_____

_____

_____

4 Complete the table by comparing the different types of sampling.

	Convenience sampling	Random sampling	Stratified sampling
How are participants sampled?			
Advantages of this method			
Disadvantages of this method			

2.3 Experimental research designs

Once a researcher has obtained a sample in the way that ensures it best represents the population, the experiment design phase commences. The experimental design that will be used within a study is an important consideration in minimising potential extraneous variables that may occur during this phase.

Experimental research may take place **between subjects**, when two different groups of participants are compared; or **within subjects**, where the same participants are used in both the control and experimental conditions.

Between subjects

There are two main types of between subjects experimental designs.

The **independent-groups design** involves randomly allocating members of the sample to either the control or the experimental group(s). Once you have your sample, you may draw participants' names randomly out of a container and assign the first name to the control group, the next name to the experimental group, and so on until all participants are assigned to groups. The independent-groups design is a quick and easy design to administer and is therefore a popular technique in experimentation. It allows large numbers of participants to be used in research in a timely and cost-effective way.

The groups in an independent-groups design should be free from bias but, due to the random nature of group assignment, there may be participant differences between the two groups. For example, members of one group may by chance be naturally more intelligent than members of another group. This design does not, therefore, effectively minimise differences in participant characteristics between the two groups.

The **matched-participants design** seeks to eradicate participant differences. It involves pairing each participant based on a certain characteristic that they share; for example, you may pair the two smartest students or the two most experienced netballers. Once you have matched these participants, you randomly allocate one to the control group and one to the experimental group. This helps to achieve an even spread of participant characteristics between the two groups, and hence minimises extraneous variables due to participant differences. It is important, however, that the matched characteristic is one that is relevant to the study and has the potential of influencing results if it was to feature more prevalently in one group over another.

One limitation of this design is that it involves a pre-test to match participants on particular characteristics (for example, you may administer an IQ test to match participants on intelligence). It is therefore more time-consuming than other designs. Also, during experimentation one participant may drop out, and in a matched-participants design this means that the other member of the pair (in the other group) must also be removed from the study.

Although a matched-participants design seeks to ensure that both groups are equal in participant characteristics, it does still use different participants who are not identical in all characteristics, abilities and motivations.

Within subjects

The **repeated-measures design** is a within subjects design. It is implemented by using only one group of participants and exposing them to both the control and experimental conditions. Because the same participants are used in the control and experimental conditions, they are therefore identical in characteristics and abilities.

This experimental design eliminates the impact of participant differences as an EV, and it also means that fewer participants can be selected for the experiment, but it does create a different problem known as **order effect**. Order effects occur when there is a change in results due to the sequence in which two tasks are completed; that is, due to the order in which participants complete the control and experimental conditions. The change in results in the second condition may be an increase in performance due to knowledge or experience in a task, or may be a decrease in performance due to boredom or fatigue after carrying out a task more than once.

For example, you may be involved in a repeated-measures study that investigates how different bathers affect swimming speed. You undertake the control condition first, in which you wear your normal bathers, and the experimental condition second, in which you wear a special swimming suit. The IV is designed to be the type of swimming suit worn, but your performance in the experimental condition may be impacted because you have already completed the control condition: it may be enhanced because you have had a warm-up swim, or it may be hindered because the first swim exhausted you. In this case, swimming performance may be influenced by this order effect, which is an extraneous variable.

One way to minimise the impact of order effects is to use **counterbalancing**. Counterbalancing involves dividing the group of participants in half and arranging the order of the conditions so that each condition occurs in a different sequence. That is, it involves exposing half of the participants to the control condition first and the experimental condition second, and exposing the other half to the experimental condition first and the control condition second. This counterbalances the potential impact of order effects on the results. In the swimming example, half of the participants would swim in their own bathers first and the new swimming suit second, while the other half would swim in the new swimming suit first and their own bathers second. Counterbalancing does not eliminate order effects occurring, but it removes the influence that order effects may have on the results.

ISBN 9780170465038

1 Complete the table by comparing the different experimental designs.

		Explanation	Advantage(s)	Disadvantage(s)
Between subjects	Independent-groups design			
	Matched-participants design			
Within subjects	Repeated-measures design			

2 You would like to investigate whether drinking an energy drink helps individuals to remember a list of 50 three-letter words. You decide to expose all your participants to a control condition, where participants will learn List A and then recall as many words as possible; and an experimental condition, where they will learn List B while drinking an energy drink and then recall as many words as possible.

a Which experimental design has been used in this experiment? Explain your answer.

b Provide two examples of order effects that may impact on the results of this study.

c Explain how you could use counterbalancing in this experiment to eliminate the impact of order effects on the results.

2.4 Placebos and procedures

While sampling and experimental design can help minimise extraneous variables, they don't prevent the impact of an individual's expectations. Participant expectations are an extraneous variable that can interfere with results because expectations can alter participant behaviour. Participants may try harder than they usually would, they could vary their responses to please an experimenter, or they may be too nervous to perform at their typical ability level. For ethical reasons, participants are mostly aware that they are being studied, but it is best they know as little as possible about the study to try to help address expectations.

Placebos and the placebo effect

To counteract the effect of participant expectations, both the control and experimental groups will typically receive some sort of treatment; however, one group – the experimental group – will receive the actual treatment (such as a new drug) and the other group – the control group – will receive a **placebo** (such as a sugar pill). A placebo is a fake or false treatment, used so that none of the participants know whether they are being exposed to the experimental condition.

Using a placebo minimises the impact of the **placebo effect** on the results. The placebo effect occurs when there is a change in a participant's behaviour due to their expectations about the treatment. For example, if the experimenter gives the experimental group a pill and does not give the control group a pill, participants in the experimental group may believe that their headache has improved, and therefore report that it has improved, simply because they received treatment. In this case the experimenter would not be able to tell whether the reported improvement is due to the effect of the drug or to the placebo effect. If participants in both the experimental and control groups receive a pill, every participant will believe that they are receiving treatment. This minimises the placebo effect because participants have equivalent expectations, which should mitigate this impact on the results.

Single-blind procedures

A **single-blind procedure** is when the participants do not know whether they have been assigned to the control or the experimental group(s). Placebos are used in this instance, and participants are unaware of whether they are receiving the placebo (control group) or the actual drug (experimental group), which reduces the impact of participant expectations on the results.

A single-blind procedure may help to balance the impact of participants' expectations on results; however, in a single-blind procedure the experimenters themselves still know which group is which. The experimenter's behaviour towards these groups (such as body language, verbal cues and preferential treatment) may therefore also influence the results of a study. This is known as the experimenter effect. It occurs when there is an unintentional change in participants' behaviour, and hence in the results, due to the experimenter's influence. For example, an experimenter may be more encouraging with the experimental group than with the control group, or may unintentionally drop hints about desired responses to help support their hypothesis. This can also be known as *demand characteristics*, where the experimenter's behaviour inadvertently causes a demand for participants to perform as the experiment predicts they will.

Double-blind procedures

To reduce the impact of the experimenter effect on the results, researchers may implement a **double-blind procedure**, in which neither the participants nor the experimenter know which participants have been allocated to the control and the experimental group(s). This, of course, must involve another person knowing which group is receiving the experimental treatment; however, this person is not directly involved with the participants, and cannot influence them.

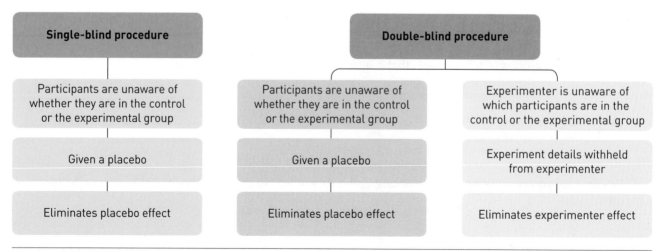

Figure 2.4 Single-blind and double-blind procedures

ISBN 9780170465038

1 Match each term with its definition.

a Double-blind procedure

i An unintentional change in participants' behaviour, and hence results, due to the experimenter's influence

b Experimenter effect

ii A fake/false drug or treatment

c Single-blind procedure

iii The participants and experimenter are unaware of who is in the control and the experimental group(s)

d Placebo effect

iv The participants are unaware of who is in the control and the experimental group(s)

e Placebo

v A change in a participant's behaviour due to their expectations of being involved in an experiment

2 Explain the similarities and differences between the terms in the table.

	Similarities	Differences
Placebo and placebo effect		
Single-blind procedure and double-blind procedure		

3 COLLECTING AND PRESENTING DATA

3.1 Data collection methodologies

In their search to answer research questions, researchers gather data in many ways. The rigour of the processes adopted is what distinguishes scientific ideas (those based on evidence and research) from non-scientific ideas (those based on opinion and assumptions). Regardless of the form in which data are collected, all fields of research require data to be collected in a systematic way. As discussed in Chapter 1, researchers must choose the methodology that best suits their research. Although much focus has been spent on the development of a controlled experiment, we learnt in Chapter 1 about scientific investigation methodologies such as case studies, fieldwork and the use of literature reviews. There are many more different types of methodologies to explore.

Observational studies

An **observational study** involves an individual observing another individual or a group of people in a natural environment, and recording observations about the behaviour they witness. The observer records behaviour they can see. Observational studies can eliminate the extraneous variable of artificiality (the effect of an unnatural environment); however, they rely on the observer's interpretation of events. This means that the observations are subject to **observer bias**, where the observer sees what they want or expect to see, which may result in a biased representation of the behaviour. Observer bias is another example of an experimenter effect.

Observational studies can take several forms. A **longitudinal study** investigates a group of people over a period of time. An example of a longitudinal study is when a researcher may visit members of the same group multiple times over a long period – sometimes decades – to gauge their responses and behaviours at different times of their lives. The *Up* series of British documentaries is a famous example of a longitudinal study. Beginning in 1964, researchers have visited the same group of people every seven years to interview them about their lives. A **cross-sectional study** investigates two or more samples of participants at a particular point in time. An example of a cross-sectional study is when a researcher looks at the difference in children's play across different cultures. Observations would be made at the same time, but in different parts of the world.

Correlational studies

A **correlational study** seeks to examine whether a relationship exists between two or more variables without the researcher manipulating any of them. A controlled experiment seeks to establish a cause-and-effect relationship (where one variable has a direct impact on another); a correlational study, on the other hand, seeks to show how two or more variables are related. It therefore does not show a causal relationship. For example, in order to investigate whether there is a relationship between stress and heart disease, a researcher may look at whether an increase in reported stress levels corresponds with an increase in the prevalence of heart disease. Similarly, opposite, or inverse, relationships can be investigated. For example, a researcher may investigate whether high levels of drug education attendance relate to lower rates of drug overdose.

Other methodologies

Methodologies take many forms and are used for a variety of purposes. Researchers may use processes around **classification and identification** when exploring new concepts, such as identifying new types of human neurons when they are discovered. Or they may invest in **product, process or system development**, where something new is designed to meet a human need. There is much advancement in this area in the use of bionic and prosthetic limbs through stimulation of the brain's electrical signals.

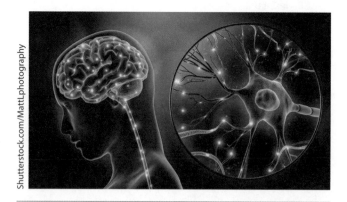

Shutterstock.com/Mattl.photography

Figure 3.1 Innovations in classification and identification may centre on newly discovered human neurons and how they function.

ISBN 9780170465038

1 Examine Figure 3.2.

   a What form of data collection is identified?

   _____

   b What is one advantage of this form of data collection?

   _____

   _____

   _____

   c What is one disadvantage of this form of data collection?

   _____

   _____

   _____

   _____

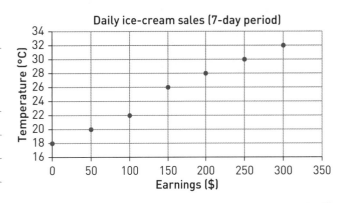

Figure 3.2

2 Examine Figure 3.3.

   a What form of data collection is identified?

   _____

   b What is one advantage of this form of data collection?

   _____

   _____

   c What is one disadvantage of this form of data collection?

   _____

   _____

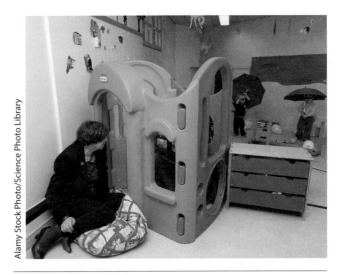

Figure 3.3

3 Examine Figure 3.4.

   a What form of data collection is identified?

   _____

   b What is one advantage of this form of data collection?

   _____

   _____

   c What is one disadvantage of this form of data collection?

   _____

   _____

Figure 3.4

3.2 Types of data

Data come in many forms and it is the researcher's responsibility to decide the form their data will take and to understand the strengths and weaknesses of each of these forms.

Primary and secondary data

Another name for data is **empirical evidence**; that is, the information psychologists gain from direct observation and measurement. If this evidence is collected through new research, such as fieldwork, observation or experimentation, it is known as **primary data**. Examples of primary data range from people's attitudes about political issues to participants' results on an intelligence or personality test. Put simply, primary data are data that a researcher gathers themselves, during a current study.

Data can also be obtained through secondary sources, such as through other people's work found in journals or academic articles. This is called **secondary data**, and is when a researcher uses previous research or knowledge in their own study. In Chapter 1 we discussed the concept of a **literature review**, where a researcher collates and analyses the findings of others, looking for correlations or patterns in the research. This is one example of a use for secondary data.

Secondary data may also be used as a basis for future predictions; to answer questions or devise theories about something that is not yet known or does not yet exist. In these cases, the secondary data obtained from past research are used as a starting point to simulate or model new data, either physically or theoretically. This is known as **data modelling**. Data modelling allows a predicted pattern to be considered so that behavioural change can happen in the present, to either prevent or mitigate the effects of something occurring.

Data modelling was particularly prominent during the COVID-19 pandemic, where known data about the virus SARS-CoV-2 and its health effects were used to predict the impact of the illness on the health system and also predict the number of potential deaths. These predictions were then used to guide subsequent governmental decisions.

Of course, techniques such as modelling and **simulation** are just predictions; the accuracy of these data is difficult to ascertain due to the large number of variables that will impact the model. There is also an inability to verify data accuracy, because the changes that are implemented as a result of the predicted data can in turn mean that the data will not be realised in the form in which they were predicted (see Figure 3.5).

Subjective vs objective data

Subjective data are based on an opinion. Such data are collected through observations of behaviour, or information based on participants' **self-reports**. Subjective data comprise personal information

Figure 3.5 During the COVID-19 pandemic, public health measures such as the mask-wearing mandate were born of secondary data used to predict the impact of the illness on health systems, if it were allowed to 'run rampant'. In turn, the full predicted impact was not realised because of the public health measures put in place.

(such as attitudes or opinions), so subjective data can be difficult to compare and analyse.

An example of subjective data would be data obtained from observing the behaviour of children in a playground. If the observer sees a child throw a ball at another child, the observer might interpret this as aggressive. However, the child who threw the ball may have delayed motor abilities and may have accidentally hit the other child with the ball; so the action looked aggressive but was, in fact, innocent. The data collected in such a situation are based only on the observer's interpretation.

Figure 3.6 Data obtained by interpreting a child's behaviour in a playground may be subjective data.

Subjective data are also obtained through self-reports, in the form of interviews, surveys or rating scales. Subjective data can provide great insight into a person's opinions and beliefs, but researchers need to be aware that such data are difficult to compare with other data. If one person says on a depression scale that they are

ISBN 9780170465038

feeling '4 out of 10' and another person says they feel '6 out of 10', the personal nature and subjectivity of the report means that we cannot assume that the individual with the lower rating is actually more depressed.

Objective data are data that can be observed and measured. Objective data are often numerical, and can be readily analysed and compared to other data. This minimises many biases encountered in research that can occur due to subjectivity. Examples of objective data are running speed, height or numbers of siblings.

Although objective data sets are easy to compare, they do not always provide the reasoning behind the data, because external factors are not taken into account. For example, in a 100-metre running race, Person A may record a faster time than Person B, and we would conclude from the objective data (time recorded) that Person A is a faster runner. However, the objective data do not account for other factors that may have contributed to Person B's result, such as injury or exhaustion.

Figure 3.7 Is the winner of the race actually the fastest runner?

Qualitative vs quantitative data

Qualitative data describe changes in the quality of behaviour, and are often expressed in words. Qualitative data can be difficult to categorise or statistically analyse because responses could take a wide variety of forms and are open to personal, observer or researcher biases. Consider a study in which a researcher wants to know participants' descriptions of a film they had seen. Every participant would describe the film in a slightly different way, would have different ways of interpreting the plot, and would have their own unique reactions to the film. This information is difficult to quantify.

Qualitative data are similar to subjective data, as they are both opinion-based. Participants are unrestricted in their responses and can provide great insight into why they feel a particular way. However, like subjective data, qualitative data are often difficult to summarise or compare with other data.

Quantitative data are data collected through systematic and controlled procedures and are usually presented in numerical or categorical form. An example of quantitative data is the number of words recalled correctly from a list, or a score on an intelligence test.

Quantitative data are similar to objective data in that they can be statistically analysed and readily measured and compared with other data. However, quantitative data can restrict participants from providing further detail and provide no insight into the reasoning behind the data that are being measured.

Mixed methods research is a very popular methodology that incorporates both quantitative and qualitative research. The findings of this type of research are strengthened because they use data obtained from quantitative measures as well as explanations or commentary obtained using qualitative measures.

Figure 3.8 Each person who sees a film will describe it in a slightly different way.

1 Study Figure 3.9, which shows the predicted rate of global temperature change if greenhouse gas emissions are reduced (blue line) and if they are not reduced (red line) between now and 2100.

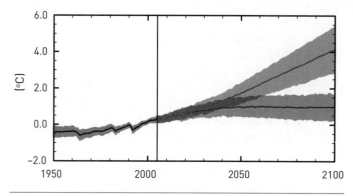

Figure 3.9 Global average surface temperature change, historical and projected

© 2022 UCAR. Adapted from Figure SPM.7, Panel (a) from IPCC, 2013 SPM in *Climate Change 2013: The Physical Science Basis. Contribution of Working Group I to the Fifth Assessment Report of the Intergovernmental Panel on Climate Change (IPCC)*.

a What type of data was used to make the predictions shown in the graph?

b List two advantages of using this type of data in any type of research.

c What might have been the researcher's purpose in creating this graph?

2 State whether the types of data are objective or subjective and qualitative or quantitative.

Type of data	Objective/Subjective	Qualitative/Quantitative
Hair colour		
Number of hours spent watching TV per day		
What people see in an inkblot test		
Descriptions of why students like doing homework		
Average height of students in a Year 10 class		

ISBN 9780170465038

3 You are interested in gathering the opinions of students at your school regarding compulsory wearing of uniforms. You don't want to restrict students' responses, so you decide to collect qualitative data.

a How could you collect these data to ensure that they are qualitative?

_____

_____

_____

_____

b What is one disadvantage of this type of data?

_____

_____

_____

c The principal would like to be able to compare students' opinions easily, so she asks you to redo your study and collect quantitative data. How could you collect quantitative data on this topic?

_____

_____

_____

_____

d You present your quantitative results to the principal, who argues that, although the data are quantitative, they are still subjective. How could that be?

_____

_____

_____

e The principal decides to make school uniforms optional but would still like to assess the students' preference for uniforms or casual dress. How could she gather objective data about this in the coming weeks?

_____

_____

_____

_____

3.3 Descriptive statistics

Once collected, data in their raw form carry little meaning or significance. To gain a picture of what the results show, data have to be organised so that they can be interpreted. **Descriptive statistics** are used to summarise, organise and describe data obtained from research.

Descriptive statistics include percentages and mathematical calculations such as measures of variance and of central tendency. Descriptive data are all around, reporting on trends, allowing comparisons to be made and telling the story of the data.

Percentages

A **percentage** illustrates the proportion of the sample that displays a particular behaviour; for example, 75 per cent of Year 12 students turn 18 in their final year of school, or 62 per cent of adolescents watch reality TV.

Using a percentage is a quick and effective way to compare the results of two or more different groups in a study. It is calculated by dividing the number of people that display a particular behaviour by the total number in the sample and multiplying the result by 100.

Measures of central tendency

Measures of central tendency involve calculations that show how typical scores, or a majority of scores, fall in a data set. There are three measures of central tendency: mean, median and mode.

The **mean** is a commonly used measure whereby all of the scores in a data set are added together and then divided by the total number of pieces of data. The mean represents the average score in a data set.

A limitation of using the mean is that it can be greatly influenced by a very large or very small score in the data set, known as an **outlier**. Outliers skew the representation of the data. For example, if you were considering the mean age of people in your Psychology class, the age of your teacher would be an outlier that could skew the results, as it would be significantly higher than the average age of the students.

The **median** is the middle number in a data set. It is calculated by arranging all of the data from smallest to largest and then selecting the piece of data in the middle. If there is an even number of pieces of data in a data set, the mean of the middle two numbers is the median. The median is commonly used when reporting VCE results, as it is not affected by outliers in the data set.

The **mode** is the final measure of central tendency and reflects the most commonly occurring number within a data set. The mode can be useful in seeing which score occurs most often, but it can be an unreliable measure for small samples.

Data set (scores on a test): 72, 73, 73, 78, 84, 85, 86, 90, 90, 90, 94, 95
Mean $= \dfrac{72+73+73+78+84+85+86+90+90+90+94+95}{12}$ $= 84.25$
Median $= \dfrac{85+86}{2}$ $= 85.5$
Mode $= 90$

Figure 3.10 Calculating measures of central tendency

Variability

Another way to describe data is by looking at how data are spread. This is known as **variability**. One measure of a data set's variability is the **range**. The range of data can be calculated by subtracting the lowest score from the highest score.

Another measure of variability is **standard deviation**. Standard deviation explores variability of data by looking at how far each individual piece of data differs, or deviates, from the mean. A low standard deviation indicates that scores are clustered around the mean and hence there is low variability in that set of scores. A high standard deviation indicates that scores are spread out from the mean and hence there is high variability in that set of scores.

Using descriptive statistics

While descriptive statistics allow data to be organised, they only describe the data and do not account for any error that may be present within the data set. For example, they can tell us which teaching method produced better scores on a test, but they cannot tell us whether the difference between the two scores is attributed to the independent variable or whether this difference occurred due to chance. In other words, descriptive statistics do not establish whether there is a cause-and-effect relationship between the variables and hence cannot be used on their own to draw conclusions from, and to be applied to, the wider population.

ISBN 9780170465038

1 Below is the number of points each of the 10 members of a basketball team scored in the final game of the season.

Player no.	1	2	3	4	5	6	7	8	9	10
No. of points	21	14	7	7	7	6	6	4	2	0

a What is the mean number of points scored by each player?

_____

b What is the median number of points scored by each player?

_____

c What is the mode for points scored by each player?

_____

d What do you believe is the best measure of central tendency to represent the point-scoring of this team? Why?

_____

_____

_____

e What percentage of the total goals did Player 1 score for her team (to the nearest whole number)?

_____

f Can we conclude that Player 1 is the team's best shooter? Explain your answer.

_____

_____

_____

_____

2 Sarah and Maeve are in two different Psychology classes at school, and they are discussing which class has done better on their recent assessment task. Both classes gained the same mean but achieved very different results. The results for each class are in the table below.

Sarah's class	21	28	13	30	7	12	26	25	18	28
Maeve's class	26	23	24	26	22	20	24	19	21	22

a Calculate the range of the scores in Sarah's class.

_____

b Calculate the range of the scores in Maeve's class.

_____

c How would you describe the variability of scores in both classes?

_____

_____

d Can you conclude that one class has performed better than the other? Explain your answer.

_____

_____

_____

_____

3.4 Visual representations of data

There are many ways to organise and present data, and some of the most common forms in psychological research are outlined below.

Frequency distribution table

A simple way of organising data, especially if you have a large amount, is to use a **frequency distribution table**. The categories being compared are placed in one column of the table. If there are several different categories, you may put them in groups, or **class intervals**, and then count how many times a piece of data fits into each interval. This is referred to as frequency.

Table 3.1 Raw scores in a hypnotic susceptibility test

55, 86, 52, 17, 61, 57, 84, 51, 16, 64, 22, 56, 25, 38, 35,
24, 54, 26, 37, 38, 52, 42, 59, 26, 21, 55, 40, 59, 25, 57,
91, 27, 38, 53, 19, 93, 25, 39, 52, 56, 66, 14, 18, 63, 59,
68, 12, 19, 62, 45, 47, 98, 88, 72, 50, 49, 96, 89, 71, 66,
50, 44, 71, 57, 90, 53, 41, 72, 56, 93, 57, 38, 55, 49, 87,
59, 36, 56, 48, 70, 33, 69, 50, 50, 60, 35, 67, 51, 50, 52,
11, 73, 46, 16, 67, 13, 71, 47, 25, 77

Table 3.2 Frequency distribution table of scores in a hypnotic susceptibility test

Class interval (scores)	Number of people in class interval
0–19	10
20–39	20
40–59	40
60–79	19
80–99	11

Graphing data

A **graph** enables large amounts of information to be neatly organised and summarised, and can show how one variable relates to another.

A **histogram** is a type of graph that can be made from the data in a frequency distribution table. Class intervals are shown on the horizontal (*x*) axis, and the frequency on the vertical (*y*) axis. Each of the bars on a histogram touches the next, because the data are continuous; that is, real values exist between each data point.

A **line graph** is any single line that connects points that relate one variable to another. A histogram can be adapted into a **frequency polygon**, which is a type of line graph that joins the midpoints of a histogram. A frequency polygon begins and ends with a point on the *x*-axis, whereas other line graphs do not necessarily start and end at zero.

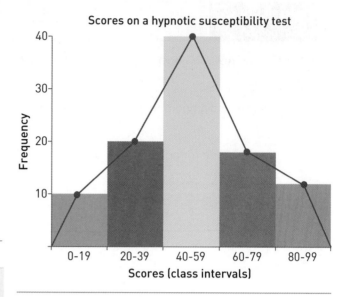

Figure 3.11 A histogram and frequency polygon

Some data that are graphed may not be continuous. For example, you may need to compare methods used for weight loss. Each method is not on a continuum of possible methods, and no two methods have a relationship to each other, so the categories you are comparing are not continuous. This is known as *discrete data*. The type of graph used to represent discrete data is a **bar graph** or **column graph**, both of which are essentially histograms in which the bars do not touch. A bar graph is presented horizontally and a column graph is presented vertically.

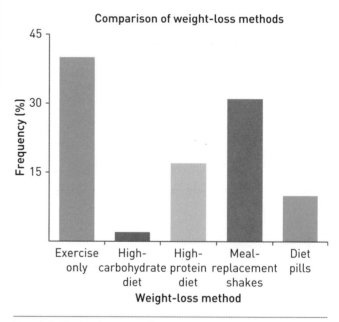

Figure 3.12 A column graph

ISBN 9780170465038

1 Sian is working as a statistician for his local hockey club. He has been assessing the average number of goals scored by each of the club's players across the last season. The raw data for average goals are recorded below.

2	1	4	3	2	0	3
8	2	0	0	1	6	4
3	0	0	11	1	3	6
2	0	3	5	1	4	2

a Complete the frequency distribution table to organise the data.

Class interval (average no. of goals)	Frequency

b Are the data represented in the frequency distribution table continuous or discrete? Explain.

_____

_____

_____

_____

c Based on your answer to part b, graph the data on the axis below in the most suitable manner. Remember to label both axes.

4 DRAWING CONCLUSIONS

4.1 Errors and uncertainty

The purpose of research is not to obtain results from the sample; it is to see whether the results obtained may be replicated in, or applied to, a broader population. To do this, considerations such as sample size and representation, the presence of extraneous variables and the strength of the results must be taken into account. Descriptive statistics only describe, organise and summarise the data; they do not allow a researcher to explore the data for error. To do this, the researcher must use **inferential statistics**.

Inferential statistics allow a researcher to make inferences about the results of an experiment; to form conclusions and to generalise findings to the population. That is, inferential statistics allow researchers to apply findings about the behaviour of small groups (samples) to the larger groups they represent (populations). To do this, inferential statistics explore whether the data obtained are a result of manipulation of the IV in an experiment or whether the results are due to chance, subject to error, or contain a reasonable amount of **uncertainty**. For example, if students in Ms Perfect's class outperformed students in Ms Lanati's class, could we simply conclude that Ms Perfect is the better teacher? Inferential statistics allow us to determine whether other factors could have contributed to the results, such as the abilities of class members, time of day the classes took place, difficulty of the class undertaken etc.

Determining the impact of these factors involves sophisticated measures of statistical calculations that look for trends, patterns and inconsistencies in the data obtained. There are many sources of errors, including **personal errors** (errors made by the researcher, such as an experimenter effect), **random errors** (errors that account for unpredictable variations or impacts on results, such as extraneous variables) or **systematic errors** (errors that occur in a consistent manner in relation to the true data value, which can be the result of something like a confounding variable). These errors can skew data and make it difficult to ascertain whether the IV has in fact impacted the DV.

If inferential statistics demonstrate that the difference in results is due to the independent variable, the results are said to be statistically significant. **Statistical significance** refers to the significance of the difference between scores; that is, whether the difference demonstrates that the results are attributed to the IV and not to chance. There will always be some error in research, but inferential statistics help to determine the level of error that is acceptable. When statistical significance is strong, there is little uncertainty in the results. Uncertainty represents a lack of knowledge of, or confidence in, the data being measured. This may be the result of missing or incomplete data, contradictory data that create additional questions, an influential extraneous variable or a source of bias. The lower the error level and the lower the probability the results occurred due to chance, the stronger the research findings.

Alamy Stock Photo/Janine Wiedel Photolibrary

Figure 4.1 When trying to compare the abilities of two teachers, the difference in student performance may be due to any number of extraneous variables unrelated to the teacher's ability.

1 Rearrange the steps in psychological research below, placing them in the correct order.

a If statistically significant, findings are applied to the broader population.

b The aim of the research is decided upon.

c Descriptive statistics are calculated.

d The experiment is conducted.

e Inferential statistics are calculated.

f Data are gathered.

2 Chase's favourite subject is maths, and he is really good at it. In the first 6 months of the year he made sure he studied for at least 1 hour a week. In the second half of the year he decided to double his study to see if it would further improve his results. His test results are below.

First half of year	Second half of year
85%	87%
82%	87%
68%	92%
88%	103%
90%	94%
Average: 82.6%	Average: 92.6%

a Using your knowledge of outliers and sources of error, explain two significant points of interest from this data set that should be considered before exploring whether conclusions can be made from these data.

b Do you believe that the average test scores hold enough statistical significance to draw the conclusion that 2 hours of study a week produce better maths results than 1 hour of study a week?

4.2 Validity and reliability

There is one more thing to consider before determining whether the results of an experiment can be applied to the broader population. One of the responsibilities of the researcher is to ensure that the methods they are using to test and assess the impact of the IV on the DV are valid and reliable. Research that does not have validity and reliability can produce false or misleading results.

Validity refers to the extent to which an assessment tool actually measures what it is designed to measure. If research claims to measure intelligence, it is only valid if it actually measures intelligence and not, for example, memory. If research claims to investigate the driving ability of teenagers, it must measure the ability of all teenagers and not, for example, only females. Valid research produces authentic results that allow researchers to examine cause-and-effect relationships between variables. The accuracy of the data is integral to ensure that the true value of the variable is being measured to produce high validity.

Reliability refers to the extent to which an assessment tool can produce results consistently, each time it is used. For example, if a person takes the same IQ test multiple times, they would achieve the same score each time if it was a reliable test. There are many ways to explore the reliability of data. If the results can be produced consistently each time, it is thought to have strong **repeatability**. If the measures that it obtains are consistent with each other, then the data shows **precision**.

Reliable data also has **reproducibility**, which is when the same results occur under changed or different conditions. For example, in a reproducible test the same results would occur at different times of day, with different participants or in different weather conditions. Research is important for advancing scientific fields and knowledge, and it is therefore important that different researchers are able to replicate or repeat the same test and make the same findings, so that the research results are strengthened.

Comparing validity and reliability

We can liken the concepts of validity and reliability to playing darts. A test with high validity and low reliability can be likened to hitting the dartboard with all the darts, but not hitting it in the same spot. The test is doing what it aims to do (results are hitting the dartboard) but the results lack accuracy.

In contrast, a test with high reliability and low validity can be likened to hitting the wall in the exact same spot with each dart, but missing the dartboard altogether. This test is not doing what it is supposed to do (results are not hitting the dartboard) but it is getting the same result repeatedly (high level of accuracy).

Ultimately, we want research to have high validity and high reliability; we want to be consistently hitting the bullseye over and over for reliable and valid research.

Testing for validity and reliability

There are many different ways to assess validity and reliability. Let's say you develop a new IQ test, and you administer both your test and an established IQ test to the same group of participants. You then compare the results from your IQ test to the results of the established IQ test. If the scores obtained from completing both tests are the same, it is likely the questions used on your test have high validity, because they appear to actually measure IQ.

To test the reliability of your new IQ test, you might have the same group of participants complete the test on a different day with a different administrator, and assess the similarity of results to those originally obtained. If the same results are obtained under different conditions, the test likely has high reliability.

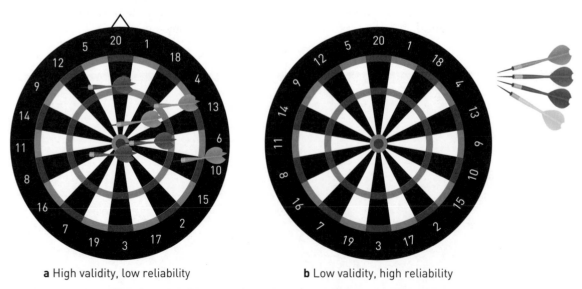

a High validity, low reliability **b** Low validity, high reliability

Figure 4.2 The dartboards illustrate a) high validity but low reliability; and b) low validity but high reliability.

ISBN 9780170465038

In the diagram below, words commonly associated with reliability have been placed around the word 'reliability'. Try to construct a similar diagram with the word 'validity'. The diagram should demonstrate your understanding of the concept of validity, and can be used to help you remember the differences between reliability and validity.

```
                                                                    R
                                                    C               E
                                                    O               P
                            S                       N               E
                            T                 S     S               A
            R               R         L   I   A   B   I   L   I   T   Y
            E               E               M       S               E
            P               N               E       T               D
            L               G                       E
            I               T                       N
            C               H                       C
            A               E                       Y
            T               N
            E
```

R E L I A B I L I T Y

V A L I D I T Y

4.3 Conclusions and generalisations

One of the main intentions of conducting research is to apply what has been learned to the population of interest. A **conclusion** is a decision or judgement about the meaningfulness of the research results. A conclusion addresses the hypothesis in research. Before making a conclusion, a researcher must consider the statistical significance of the results, the sample size and sources of potential error. It is important that the researcher is confident that the change in the dependent variable is due to the independent variable within the stated population.

If a conclusion has been made and the research meets certain other criteria (see below), the research findings can be applied to the broader population, or the wider group of research interest. When we apply research findings to the wider population, this is known as a **generalisation**.

A generalisation should only be made if all of the following criteria are met.

- The results are statistically significant.
- The sample is representative of the population.
- The method of sampling is appropriate.
- Wherever possible, extraneous and confounding variables have been controlled for.
- There are no known sources of error or bias.

For example, a researcher may find a statistically significant link between listening to music and enhancement of memory, but if the sample only contained 16-year-olds, the link cannot be generalised to other age groups. As another example, a teacher may have conducted research investigating the effect that the number of hours of study had on scores obtained in a maths test; however, if her sample only consisted of students in her class, this sample is biased and not representative of the population. A final

example might involve investigating the influence of memory aids on the capacity of short-term memory. If Group A, which uses mnemonic devices, comprises only males, and Group B, which does not use mnemonic devices, comprises only females, then gender is a possible confounding variable and the findings cannot be generalised.

Generalising results to the population is one of the main goals of psychological research. Experimenters do not often endeavour to discover something for only a small group of people; instead, they seek to discover something that can be applied to a large group of people, or to an entire population.

www.CartoonStock.com/Marty Bucella

I didn't just jump to conclusions. I hopped and skipped first.

Figure 4.3 Jumping to conclusions, without appropriate experimental procedures, can create misleading findings.

ISBN 9780170465038

1 Read the following articles. Discuss why it is difficult to generalise the results for each of these studies.

a Does drinking coffee help you live longer?

There's only one thing better than a hot cup of coffee in the morning: a new research paper [published in *JAMA Internal Medicine*] telling you your daily habit is good for your health.

Like many previous studies, the JAMA paper found people who drank coffee had a lower risk of dying of any cause ... over the course of the study.

This was a prospective trial, which tracked almost half a million British residents over ten years ... In a baseline questionnaire, subjects gave detailed responses to coffee consumption (how much, how often, what types of coffee and whether it was caffeinated or decaffeinated), as well as other factors such as alcohol, tea, race, education, physical activity, body mass index and smoking. The participants' health status was monitored during the study and, if they died, their cause of death was determined by the National Health Service using internationally recognised criteria.

After taking into account factors such as smoking and alcohol intake, the researchers found ... coffee drinkers were around 5–10% less likely to die from heart disease, cancer and other causes during the study period than non-coffee drinkers.

Compared to non-coffee drinkers, those who consumed one cup of coffee a day had an 8% lower risk of premature death; this increased to a 16% lower risk for those who drank six cups a day. People who drank up to eight cups of coffee a day were 14% less likely to die prematurely than non-coffee drinkers. This pattern was seen for all types of coffee, including instant and decaffeinated coffee.

As with the previous studies, this is a correlation study. So, while there was an association between coffee consumption and a lower risk of death, we still can't say coffee was the cause of the lower risk of death.

There may be some other environmental variable that was not accounted for. Coffee consumption may entail more walking, for instance, which was not captured in the lifestyle questionnaires.

Source: Musgrave, I. (2018, July 5). *Research Check: does drinking coffee help you live longer?* Retrieved March 2022 from https://theconversation.com/research-check-does-drinking-coffee-help-you-live-longer-99287

b Yoga can help keep expectant mothers stress free

For the first time, researchers in the UK have studied the effects of yoga on pregnant women, and found that it can reduce the risk of them developing anxiety and depression. Stress during pregnancy has been linked to premature birth, low birth weight and increased developmental and behavioural problems in the child as a toddler and adolescent, as well as later mental health problems in the mother. A high level of anxiety during pregnancy is linked with postnatal depression, which in turn is associated with increased risk of developing depression later in life.

In a paper published in the journal *Depression and Anxiety*, academics from Manchester and Newcastle Universities in the UK show that women who attended a yoga class a week for eight weeks had decreased anxiety scores compared to the control group who received normal antenatal treatment.

The study was carried out in Greater Manchester and looked at 59 women who were pregnant for the first time and asked them to self-report their emotional state. A single session of yoga was found to reduce self-reported anxiety by one third and stress hormone levels by 14%. Encouragingly, similar findings were made at both the first and final session of the eight-week intervention.

Source: *First evidence that yoga can help keep expectant mothers stress free*, The University of Manchester (1 May 2014), https://www.manchester.ac.uk/discover/news/first-evidence-that-yoga-can-help-keep-expectant-mothers-stress-free/

5 ETHICS

5.1 Ethical concepts – experimenter considerations

The term **ethics** refers to the moral principles and codes of behaviour that apply to all psychologists, regardless of the field in which they work. The Australian Psychological Society (APS) publishes a *Code of Ethics* that outlines the general ethical principles that govern the behaviour of psychologists. The APS also publishes the complementary *Ethical Guidelines*, which applies the *Code of Ethics* to situations encountered in everyday professional practice. *Ethical Guidelines* is regularly updated. The *Code of Ethics* can be viewed at the APS website (psychology.org.au).

Interestingly, over the years, ethical guidelines governing psychological research have become stricter, so that some psychological experiments conducted in the past would be considered unethical by today's standards. In the documentary *Three Identical Strangers*, triplets are separated at birth solely for the purpose of experimentation. This would never be allowed by today's standards. Similarly, Zimbardo's famous 'Stanford prison experiment' and Milgram's 'electric shock' experiments would not be approved today from an ethics standpoint. The more we know about human behaviour, the more informed the decisions of ethical committees become, ensuring we provide the greatest level of protection to those in a researcher's care.

Stanley Milgram Papers (MS 1406). Manuscripts and Archives, Yale University Library. Reproduced with permission from Michele Sara Marques.

Figure 5.1 Stanley Milgram's 'electric shock' experiments of the 1960s would not pass an ethics committee today.

Prior to the commencement of any research, the researcher submits a research plan to an **ethics committee** for approval. The ethics committee may be made up of academics and professionals with an understanding of the impact of research on an individual's health and wellbeing; this ensures that the participants' welfare is considered.

The ethics committee and researcher must investigate the potential benefits of the research to society, which need to be weighed against the potential risks or discomfort to participants. This ethical consideration is known as **beneficence**, where research is considered through the scope of maximising the benefits to society while minimising harm to others. Where potential harm may be involved, either physically or psychologically, it is important to consider **non-maleficence**. In the medical profession, the principle of non-maleficence involves avoiding causing any kind of harm at all. In scientific research, the principle of non-maleficence implies that, if potential harm/discomfort is involved, the harm/discomfort does not outweigh the potential benefit(s) that could eventuate. For example, is it worth causing psychological distress to infants if we can discover guidelines for parental attachment? Examples of weighing beneficence and non-maleficence are prevalent throughout the medical world.

When research commences, it is important that the researcher shows **respect** for all involved. Respect in research is shown through the consideration of an individual's welfare, but also through the appreciation of their own uniqueness, autonomy and freedom of expression. It is the hope of research to discover things that can create an impact or improve quality of life, and therefore being selected for research itself can be a process of careful consideration. Let's say a researcher has discovered a new treatment to stop the progression of Alzheimer's disease. The trial lasts for 1 year and, if successful, could give families more coherent time with their loved ones. How do we decide who is in the trial (experimental group) and not (control group)? This is where the ethical principle of **justice** is involved. Justice ensures fair treatment for all, from selection through to result collection.

One final ethical consideration expected from the experimenter is that they act with **integrity**. Integrity involves the researcher's commitment to the honest conducting and reporting of research. Integrity also relies upon scrutiny of the research and of its procedures so that the knowledge and understanding gained through research can be trusted in its broader application.

The work of ethics committees is integral in governing research because of the broader implications of the findings. Research findings can be used to inform policy, to distribute funding and to shape our behaviours. Potential research must therefore be scrutinised to best understand why the research is being conducted and who the findings stand to benefit.

1 Imagine you are part of an ethics committee. Write at least three questions you would ask the researcher in each of these scenarios.

a A researcher has submitted a proposal to separate infants from their parents, for two weeks, when each infant is 5 months old. They are seeking 100 infants who will live in the care of a registered nurse for the two-week period. The researcher will record the behaviour of the infants during separation and when re-introduced to their parents. The researcher wants to investigate the impact of separation from parents to inform research on the impact of childcare during the first 6 months of life.

b A new medication for depression has been found to be effective in small clinical trials. A researcher is seeking approval to use the new treatment on Year 12s during their final year of study to see whether it continues to show effectiveness. One thousand subjects between 16 and 18 years old are sought for the trial. Half will be given the new medication that reduces the experience of depression and enhances focus and concentration; the other half will be given a placebo treatment.

2 Use the space below to write simplified definitions for the ethical principles discussed so far. Ensure that you understand the terminology used in a way that makes sense to you.

5.2 Ethical guidelines – participants' rights

Once research is approved by an ethics committee, it is important that participants in research are protected. The researcher should adhere to any relevant ethical guidelines and ensure that the following six participant rights are adhered to.

Voluntary participation ensures that a participant willingly decides to take part in an experiment. Participants must not experience any pressure or coercion to participate, nor be threatened with any negative consequences if they decide not to participate in an experiment.

Informed consent needs to be obtained before an experiment commences. The researcher must obtain written permission from each participant in a study, stating that they consent to participating in the study and have been given all necessary information. The consent form must inform the participants about their rights, as well as any possible physical or psychological harm that may be encountered during the experiment. Where it is possible and reasonable, participants must be informed about the research procedures employed in the study. If a participant is under the age of 18, or is legally unable to give consent, the participant's parent or guardian should complete the consent form.

Withdrawal rights refers to the right of the participant to cease their participation in a study at any time without negative consequences or pressure to continue. This guideline must be adhered to during an experiment and also after an experiment; if a person feels uncomfortable during any follow-up activities they are involved in, or wishes to remove their results from being used in the study, withdrawal rights ensure that they can do this without consequence.

Confidentiality is a participant's right to privacy with regard to access, storage and disposal of information collected about them that is related to research. A participant's involvement in, and results from, an experiment cannot be disclosed to anyone else unless written consent has been obtained.

Deception in research should not occur unless it is necessary. It is used in some cases where giving participants information about an experiment beforehand might influence their behaviour during the study and thus affect the accuracy of results. However, deception in research must be used with caution and, when it is used, researchers must ensure that all participants are thoroughly debriefed.

Debriefing involves participants being informed of the study's true purpose once the experiment has ended. During debriefing, a researcher must also correct any mistaken attitudes or beliefs held by the participants, and explain all deception related to the conducting of the experiment. The experimenter must also provide an opportunity for the participants to gain access to information about the study, including procedures, results and conclusions, and provide access to additional support through counselling, as required.

Shutterstock.com/nito

Figure 5.2 A researcher must obtain informed consent from every participant in a study.

ISBN 9780170465038

A university lecturer has been asked to conduct research investigating how mild pain affects memory formation. She decides to use students from her tutorial class. She separates the students into two groups. She shows students in Group A a list of 20 words and then gives them 2 minutes to recall as many of the words as possible. She shows students in Group B the same list of 20 words, administers a mild electric shock to them, then gives them 2 minutes to recall as many of the words as possible.

1 What are two major issues that an ethics committee may have with this research? Explain why they are areas of concern.

_____

_____

_____

2 Explain how you would address each of the following ethical guidelines if you were conducting this research.

  a Voluntary participation

_____

_____

  b Informed consent

_____

_____

  c Deception

_____

_____

  d Withdrawal rights

_____

_____

  e Debriefing

_____

_____

  f Confidentiality

_____

_____

3 Describe one famous past psychological experiment or piece of research that you have discussed in class (not mentioned in this book) that you would consider to be unethical by today's standards. Explain why you consider it to be unethical.

_____

_____

_____

_____

6.1 Writing a scientific poster: Title and Introduction

Throughout your studies of psychology, whether at school, university or beyond, you will have to report on findings from research. Some scientific reports involve you reporting from secondary data, and others from primary data. Some scientific reports are extensive, while others are summaries. This chapter includes sections that may feature in scientific reports or posters in various forms, and showcases some of the features under each subheading.

All scientific reports, regardless of style, are to be written in a formal scientific manner. Therefore, you must make sure you avoid words such as 'I' or 'we' and that you write in past tense. Your report is explaining what you did, not what you are going to do.

Title

The **Title** of your report should be a description of the variables that were manipulated and measured in the investigation. A good formula for working out the Title is to use the following phrase: 'The effect of [the IV] on [the DV]'. Alternatively, you may use the question under investigation as your Title.

Introduction

The **Introduction** is important because this is where you introduce your topic. Think of the opening of the Introduction as a funnel, starting very wide (broad) at the top and then becoming very narrow (specific) at the bottom. Introduce your topic as you would in an essay. What are the definitions of the key terms you are referring to? Is there background knowledge that needs explaining? Ensure that these concepts are linked together in a logical, sequential manner.

Once you have provided a funnel of background information, explain the reason that this investigation is now being embarked on. You may refer to past research in this field demonstrating what others have found before you. Remember to cite past publications correctly and include them in your reference list.

To finish the Introduction, you will need to state the aim (a statement that explains what you are intending to investigate in an experiment), hypothesis(es) and the independent and dependent variables.

To the right is an example of a Title and an Introduction.

The effect of alcohol consumption on driving ability

Introduction

During day-to-day life, an individual can move between different states of consciousness; from normal waking consciousness (NWC) to altered states of consciousness (ASC). When in NWC our thoughts are logical and we are aware of our surroundings. When in an ASC we can experience perceptions and thoughts that are less logical, a decrease in our inhibitions and decreased awareness of our surroundings. Drinking alcohol can make individuals enter an ASC. Due to the effects that can occur in an ASC, there are strict laws in Australia that prevent people from driving a car while under the influence of alcohol. Guidelines state that it is legal to drink alcohol and drive if you have a blood alcohol level less than 0.05 per cent for a fully licensed driver. But is the effect of even that amount dangerous?

Previous research suggests there is a strong relationship between alcohol consumption and altered driving ability. Smith (2020) found that, when participants consumed more than two alcoholic drinks in 1 hour (and fewer for some women), their driving ability was impaired. He found that participants who drank more than two alcoholic drinks in 1 hour were five times more likely to have a car accident and 15 times more likely to be at fault than people who had fewer than two alcoholic drinks in 1 hour.

The aim of this study was to replicate Smith's findings by examining the effects of alcohol consumption on driving ability when using a simulator. It was hypothesised that Victorian university students who consumed two alcoholic drinks in 1 hour would have more errors on a driving simulator than those who had two non-alcoholic drinks in 1 hour. The independent variable was the number of alcoholic drinks consumed. The dependent variable was the number of driving errors made on a driving simulator.

ISBN 9780170465038

1 Write a Title for each of the following research experiments by using the format 'An investigation to test the effect of [the IV] on [the DV]'.

a A scientist wants to investigate how to increase serotonin levels to help combat depression. He decides to test a new serotonin medication on a random sample of university students, and compares the effects with students who have had no medication.

_____

_____

b An educational psychologist is trying to see whether or not confidence in public speaking can be increased using a variety of preparation techniques. She teaches one group of students how to write a good speech, and has another group take acting classes.

_____

_____

2 Write the beginning of an Introduction that presents relevant psychological concepts for a piece of research that explores the impact of examinations on a young person's wellbeing.

_____

_____

_____

_____

_____

_____

_____

_____

3 Number each of the sections below in the order they appear in the Introduction section of a research report.

| ☐ | Independent and dependent variables | ☐ | Hypothesis |

| ☐ | Past research | ☐ | Theories and definitions |

| ☐ | Aim |

6.2 Writing a scientific poster: Methodology and method and Results

Methodology and method

The **Methodology and method** section can typically be broken into subsections and should include the identification of any risks, and how they were managed, wherever relevant. You should first state the **methodology** you used to answer your research question.

The subsections should then outline the **method** you used to conduct the investigation. The first subsection should be 'Participants'. In this subsection you should give the details of the sample, including how many participants were used in the study and their ages and genders. You might also mention the sampling method used.

Next is the 'Materials' subsection. In this subsection you must list every item that was required to conduct the study. You should also mention any question sheets or surveys you used, and ensure that you remember to attach these to the back of your report as an **Appendix** (we will discuss the Appendix in Section 6.3).

The final subsection is the 'Procedure'. This section is preferably completed using dot points and should be a step-by-step outline of how the experiment was conducted. Think of this section as though it were a recipe: if someone wanted to replicate your experiment, they would simply follow the steps in your procedure. This section should begin by explaining the participant selection process and end at the point when results were collated. Ensure that you provide detail as to the way you gathered and analysed your data.

Results

The **Results** section should feature a visual representation of the primary data you have collated, or the secondary data you have sourced. Make sure the Results represent collated/sorted data – that is, the descriptive statistics – and not raw data. Raw data comprise all of the individual participants' results, and are too long to be included in the Results section. Attach the raw data to the back of your report as an Appendix.

The Results section may include graphs or tables, but you should only graph the average or the percentage, or whichever data presentation method you believe best illustrates your findings. Your visual representation must include a title and must be clearly labelled. Make sure that you label the x- and y-axes if you are using a graph.

Under your visual representation, explain your findings in words. There is no need to elaborate or explain the Results in this section; simply state what the findings were, highlighting any trends, patterns or relationships.

Opposite is an example of a Methodology and method and a Results section.

ISBN 9780170465038

The effect of alcohol consumption on driving ability

Methodology and method

A controlled experiment was used to investigate the research question.

Participants

One hundred first-year university students from Melbourne University were used in the study. Ages ranged from 20 to 40 years, with equal numbers of males and females.

Materials

The following materials were used to conduct the experiment:

- alcohol
- a driving simulator
- drinking glasses.

Procedure

- Participants were selected via random sampling and randomly allocated into either the experimental or the control group.
- Participants in the experimental group consumed two alcoholic drinks in 1 hour.
- Participants in the control group consumed two non-alcoholic drinks in 1 hour.
- After the allocated time, both groups of participants used the driving simulator.
- The number of errors was counted for each participant and used to measure their driving skills. The average scores were then calculated for each group.

Results

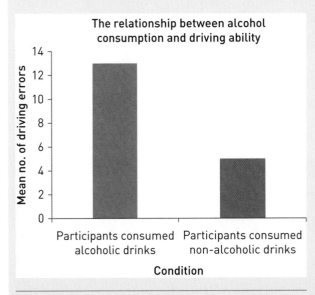

Figure 1

The results are shown in Figure 1. The mean number of driving errors made by the control group was 5. The mean number of driving errors made by the experimental group was 13. The results show that the experimental group made more driving errors on the simulator than did the control group. The raw data can be found in Appendix 1. The results were found to be statistically significant.

The Methodology and method and Results sections of the research investigation below include many mistakes. List all errors you can find with this research report and explain in detail why they are incorrect.

Methodology and method

Materials

In this experiment, I used a worksheet entitled 'How to write speeches' (attached in the References section). I also used a drama classroom, a lecture hall for public speaking and a confidence survey that the participants completed and handed in (individual participant responses are known as rare data and are attached to the References section).

Participants

In this experiment I used 20 participants in one group and 20 in the other.

Procedure

The participants were selected from a local primary school and two classes were chosen by convenience sample. One group of participants took a series of three drama classes. The others did not. All students then gave a speech in a lecture theatre and rated on a rating scale how confident they felt out of 10. These results were then collated. After this, a research report was written.

Results

These results show all of the scores that the participants gave in terms of their rating of confidence.

The confidence ratings for participants who took the acting classes (out of 10):

6	9	7	8	9	7	6	3	9	10
8	9	8	7	5	10	9	9	8	9

The confidence ratings for participants (out of 10):

6	9	7	8	9	7	6	3	9	10
8	9	8	7	5	10	9	9	8	9

Writing a scientific poster: Discussion, Conclusion, References and Appendices

Discussion

You should begin the **Discussion** by addressing your hypothesis and stating whether your results refute or support it. Make sure you provide sufficient evidence as to why your hypothesis is supported or rejected by analysing and evaluating the data. Then, compare your findings with past research, or psychological concepts discussed in the Introduction, and explain any differences in the results, or anything unexpected.

You must then discuss extraneous variables and the effect they may have had on the results, including any limitations of your methodology or of the data-collection process. Suggestions for future improvements to the research should appear in this section as well. Round out this section by making possible suggestions for future investigations.

Conclusion

Next, you must provide a **Conclusion**, which is a statement that addresses the original aim or research question. You may highlight the extent to which the research has addressed the research questions; however, you should not introduce new material. You may be in a position to suggest whether the results found in this study could be generalised to the population from which the sample was drawn. This may or may not be possible, based on a range of experimental conditions discussed in Chapter 4.

When completing a scientific poster, your communication statement is drawn from your conclusion and is a one-sentence summary of the key finding of the investigation.

References

Throughout your research investigation, you should cite any publications that you refer to, or that you paraphrase. The **References**, therefore, is a list of all publications you have cited in your research investigation. All references should be cited according to the American Psychological Association (APA) format. The format for books and journal articles is shown below, but there are APA conventions that should be followed for all types of publications.

For books, the format is:

Surname, initial. (Year). *Title of book*. City of publication: Publisher.

For journal articles, the format is:

Surname, initial. (Year). Title of the article. *Title of Journal*, *Volume of Journal* (edition number), page numbers.

Appendices

An Appendix (plural: appendices) is something that is supplementary to the main body of a scientific report and should appear at the end of the report. Appendices would include such things as the raw data from the results, as well as any sheets, questions or surveys used in the investigation. Appendices should be numbered and referred to where appropriate in the body of the report.

When completing a scientific poster in your studies, you will not attach Appendices. Instead, your raw data and any other supplementary material will be contained in your logbook, which will be submitted for assessment.

Over the page is an example of a Discussion, a Conclusion and References.

The effect of alcohol consumption on driving ability

Discussion

It was hypothesised that Victorian university students who consumed two alcoholic drinks in one hour would make more errors on a driving simulator than students who consumed two non-alcoholic drinks. The hypothesis was supported, in that the results suggest that participants who consumed alcohol made more errors on the driving simulator than the participants who did not consume alcohol.

The findings also support previous research by Smith (2020), who found that alcohol consumption is likely to cause impairment in driving ability.

Extraneous variables that could have affected the results include individual differences in tolerance to alcohol. Some individuals can drink more than others and retain their driving ability. Additionally, factors such as mood, weight, body fat content and whether a person has just eaten all affect how much an individual could be affected by alcohol. The driving ability of participants was also not tested beforehand, so skill could have been a factor in the results obtained. One way to overcome these individual differences is to employ a repeated-measures design. Instead of having two independent groups of participants (the experimental and the control group), the same subjects could be used in the experimental and control groups.

Another extraneous variable may be that of artificiality, as participants were using a driving simulator and hence the situation was not real. They were also aware that they were being watched as they used the simulator, which may have resulted in participants being more careful than they usually would when driving, or they may have felt nervous. This may have led to an increase or decrease in the number of driving errors due to the simulated environment. One way to overcome this variable is to conduct more trials, and to ensure that the experimenter is not in the room. After conducting more driving trials, the participant may become more relaxed and familiar with the artificial environment, and hence behave in a naturalistic way.

Conclusion

It can be concluded that consumption of alcohol increased the number of driving errors, and hence consumption of alcohol has a negative effect on driving ability in first-year students from Melbourne University. These results are unable to be generalised to all Victorian university students because the sample was drawn from only one university.

References

Smith, J. (2020). The effect of alcohol consumption on driving ability in doctors. *Journal of Traffic Accidents, 12*, pp. 44–56.

ISBN 9780170465038

1 Students were required to do an oral presentation as part of their English curriculum. They were interested to find out whether rehearsing their presentation via rote learning was more effective than learning by using memory aids. To investigate this research question, 50 students learned their speech by rote and another 50 used memory aids. Their scores on the oral presentation were then compared. Discuss two extraneous variables that may impact this experiment, and how they could be avoided in the future.

Extraneous variable:

_____

Future improvements:

_____

_____

Extraneous variable:

_____

Future improvements:

_____

_____

2 Reference the two books below using APA format. Use underlining to represent italics.

a Brain Surgery for Beginners, written by Jean Geoffs. Published by Ebony House in Melbourne in 2018.

_____

_____

_____

b Measuring your Emotional Intelligence, published by Construction Place in Canberra. Written in 2021 by Thomas Peters.

_____

_____

_____

3 Looking at the research scenario in Question 1, describe all of the resources that may have appeared in the appendices for this study.

_____

_____

_____

_____

_____

_____

_____

_____

6.4 Prepare your own scientific poster

Use this section to plan for, and conduct, a small experiment. This format could be used as a logbook for any future research as well. Once you have completed the experiment, present your findings in a scientific poster report.

Below is some background information to help you with your Introduction, but try to consult other sources to flesh out this section.

The Atkinson–Shiffrin model of memory

The Atkinson–Shiffrin model of memory is separated into three subsystems: sensory memory, short-term memory and long-term memory. It is the process of encoding information from short-term to long-term memory that is of particular interest to psychologists and educators alike. Once a memory is consolidated in long-term memory, it is relatively permanent. This means that once you transfer material learned in school to your long-term memory, it will stay there indefinitely.

There are always new theories and ideas emerging as to the best ways to enhance memory. It is widely known, for example, that aids such as mnemonic devices can help to enhance memory. Additionally, many students claim that listening to music helps them to retain information; others say that writing out definitions over and over again helps them to retain information.

What are the benefits of chewing gum?

Do you have a habit of chewing gum? You may think of gum chewing as something that gets you in trouble in a school classroom, or something that drives your parents crazy. Well, you may be vindicated! Research suggests that chewing gum regularly may be a natural way to boost concentration.

It is known that chewing gum is used for combating sleepiness during work, and while learning and driving, which suggests a link between chewing and maintaining memory and attention (Hirano & Onozuka, 2014). Studies by Hirano et al. (2008) examining the effects of chewing on neuronal activities in the brain during a working memory task showed that chewing may accelerate the process of working memory. Allen and Smith (2015) found that

chewing gum during the workday was associated with higher productivity.

Scientists theorise that the act of chewing stimulates heart rate, making the brain more active. Also, chewing stimulates the pancreatic hormone insulin, which can improve memory. So, go on, try chewing gum during times when you need your memory to be at full throttle, such as during a test or an oral exam.

Allen, A. P. & Smith, A. P. (2015). Chewing gum: Cognitive performance, mood, well-being, and associated physiology. *BioMed Research International.* http://dx.doi.org/10.1155/2015/654806

Hirano, Y., Obata, T., Kashikura, K., Nonaka, H., Tachibana, A., Ikehira, H. & Onozuka, M. (2008). Effects of chewing in working memory processing. *Neuroscience Letters, 436*(2), pp. 189–92.

Hirano, Y. & Onozuka, M. (2014). Chewing and cognitive function. *Brain Nerve, 66*(1), pp. 25–32.

ISBN 9780170465038

After reading some psychological theory and information about past research, it is now your turn to determine whether chewing can enhance your memory.

For this experiment you will need:

- participants
- a list of **15** words (they could be nonsense words or three-letter words)
- something for participants to chew.

Conduct an experiment in which participants learn and then attempt to recall the list of words. Half of your participants are to chew as they do so; the other half are not to chew.

You are the experimenter and the rest of the details are up to you. How many people will you use? Will there be a time limit for recall? Will participants chew lollies, gum or something else?

Follow the steps below to prepare for your experiment and gather your data.

1 Start by writing down what it is you are aiming to find.

   a The aim of this experiment is to:

   b The independent variable is:

   c The dependent variable is:

2 Now that you know what you are investigating, you will need to formulate a hypothesis. It is hypothesised that:

3 You now need to design a method. Answer the following questions to help you finalise the details of your method.

   a How many people will you test?

   b Where will you gather the participants? How?

   c List all the materials you will need to conduct the experiment. When you have listed them, gather them in preparation for using them in step 6.

4 Now, write up your step-by-step procedure (remembering to use past tense when writing formally).

5 You also need to consider all elements that may become extraneous variables. It is important to eliminate as many of these as possible, to help obtain reliable results. List possible extraneous variables that may arise and state how you will control for these.

_____

_____

_____

_____

_____

_____

_____

6 Now you are ready to conduct your experiment and collect your data.

a Record the details of your participants in the space below. Include number, age and gender of participants.

_____

_____

_____

_____

b Collect your raw data and fill in the table with it.

	Group A: Chewing	Group B: Not chewing
Number of words recalled		

7 It is now time to use descriptive statistics to organise your data. You could use percentages, means, medians or modes; however, a mean is probably the best measure for this study.

	Group A: Chewing	Group B: Not chewing
Number of words recalled		

8 Has your hypothesis been supported or refuted? Was this what the past research suggested?

_____

_____

_____

_____

9 It is now time to prepare to write your scientific report. Check that you have all the necessary materials before you start writing.

Research investigation checklist	
Number, ages and genders of participants	
List of all materials used	
Raw data for appendices	
Past research	
All referencing details	

ISBN 9780170465038

Poster report format

Title: _____

Student name: _____

Introduction:

Communication statement (this should report the key findings of the investigation as a one-sentence summary)

Discussion:

Methodology and method:

Conclusion:

Results:

References and acknowledgements:

7 EXAMINATION SKILLS

7.1 Examination questions

Report writing is only one way that knowledge of key science skills can be demonstrated. You can also show your knowledge of these skills in the form of multiple-choice, short-answer or extended-answer questions, as you will in the Psychology examination. Below are some sample questions on the research methodologies you have studied in this workbook.

1 In a recent study in England, pregnant women who smoked were recruited to a study that aimed at stopping their smoking during pregnancy. Half of the pregnant women in the study were offered one-on-one interviews and nicotine patches to support their efforts to quit; the other half were offered the support received by the first group as well as shopping vouchers every few weeks up the value of $400 if they abstained from cigarette smoking. The results of the study showed that only 9% of women in the control group stopped smoking compared with 20% of women in the shopping voucher group.

a Identify the IV and DV and write a possible hypothesis that could have been investigated in this study.

(4 marks)

_____

_____

_____

_____

_____

_____

_____

_____

b One aim of research is to minimise possible extraneous variables. Identify and explain a sampling technique and an experimental design that the researchers of this study may have used, in their efforts to minimise bias and/or the effects of extraneous variables. In your discussion, explain the extraneous variables that the chosen sampling technique and experimental design would seek to minimise.

(6 marks)

_____

_____

_____

_____

_____

_____

_____

_____

_____

_____

_____

_____

_____

c Discuss any ethical concerns that may be relevant to the research. (4 marks)

_____

_____

_____

_____

_____

_____

_____

d Could these findings be generalised to Australian pregnant women? Explain your answer. (2 marks)

_____

_____

_____

_____

2 With more and more Victorian schools making computers compulsory for students in the classroom, there is a push from the Education Department to have end-of-year examinations completed using technology. One factor currently under investigation is whether or not using computers for exams will allow students to obtain better results than they do currently, because the ability to type (rather than write) answers should lead to better legibility and increased speed. A trial group was set up at two secondary schools with students of comparable academic ability. Both schools had been using compulsory electronic devices throughout the year for their normal classroom learning. Students at Secondary School A sat the end-of-year examination on an electronic device, whereas students at Secondary School B completed the same examination but on paper, and hence completed the traditional written examination.

Two dependent variables were examined: speed of completion and performance on the exam.

a Write two hypotheses for this research (one for each DV). (6 marks)

_____

_____

_____

_____

_____

_____

_____

_____

b Write an advantage and a disadvantage of using a matched-participants design for this research. (2 marks)

_____

_____

_____

_____

_____

_____

_____

c Discuss at least two extraneous variables in this research. (4 marks)

_____

_____

_____

_____

_____

_____

_____

_____

d Discuss at least one ethical consideration in this research. (2 marks)

_____

_____

_____

_____

_____

_____

_____

_____

3 Entering university can be a daunting time for any new student. One of the most difficult things that students must adapt to is the change from traditional classes to lectures. Note-taking becomes a crucial skill; students must be able to summarise and organise content.

Three hundred first-year university students undertook a 3-week course in note-taking skills before commencing university. Another 300 first-year students undertook a 3-week course on getting to know university resources, which featured tours of the libraries on campus. The students who took the note-taking skills course performed, on average, 5 per cent higher on the first-semester examinations than those students who did not take the note-taking course.

a Write a possible hypothesis for this study and identify the independent and dependent variables. (5 marks)

_____

_____

_____

_____

_____

_____

_____

_____

_____

ISBN 9780170465038

b Show your understanding of research principles by explaining why two courses were offered in this
research and what the most appropriate type of descriptive statistic may be to present these data.　(4 marks)

_____

_____

_____

_____

_____

_____

_____

_____

_____

_____

_____

_____

_____

_____

c Show your understanding of components of a discussion by outlining two extraneous variables
and writing a possible conclusion for this study.　(5 marks)

_____

_____

_____

_____

_____

_____

_____

_____

_____

_____

_____

_____

_____

_____

_____

GLOSSARY

Accuracy
A lack of error or bias in a measure or score

Aim
A statement that explains what you are intending to investigate in an experiment or research investigation

Appendix (plural: appendices)
In a scientific report or poster, the section that includes any raw data used in the experiment, as well as additional materials such as questionnaires

Artificiality
An extraneous variable whereby the unnatural environment in which an experiment is conducted impacts on participants' behaviour

Bar graph
A graph of discrete data presented horizontally; its bars do not touch

Beneficence
An ethical principle where research is considered through the scope of maximising the benefits to society while minimising the harm to others

Between subjects (design)
An experimental research design in which two different groups of participants are compared; each group is assigned to a different condition and experiences only one of the experimental conditions

Case study
An in-depth or detailed study of a particular person, activity, behaviour or event

Causal
Implying that one variable created a change in another; there is a cause-and-effect relationship

Class interval
A category or group of data, used when sorting raw data into descriptive statistics

Classification and identification
The arrangement and recognition of phenomena into a particular set

Column graph
A graph of discrete data presented vertically; its bars do not touch

Conclusion
A decision or judgement about the meaningfulness of the results of a research investigation

Confidentiality
An ethical guideline that describes a participant's right to privacy with regard to access, storage and disposal of information about them that is related to a research study

Confounding variable
A variable other than the independent variable that causes a change in the dependent variable, and whose effects may therefore be confused with those of the independent variable in a study

Control group
The group in an experiment that is exposed to the control condition, in which the variable under investigation is absent

Controlled experiment
A study under controlled conditions that investigates a cause-and-effect relationship between two or more variables, and tests a hypothesis

Controlled variable
Any variable that is constant in research conditions

Convenience sampling
A sampling technique used in selecting participants for a study, which involves selection of participants based on easy accessibility and availability

Correlational study
A data-collection technique that involves determining a relationship between two or more variables without the researcher manipulating any of them

Counterbalancing
A technique used in a repeated-measures experimental design, which involves arranging the order of the conditions so that each condition occurs equally often in each position

Cross-sectional study
When a researcher seeks to investigate two or more samples of participants at a particular time

Data modelling
The construction and/or manipulation of either a physical or a conceptual model that represents a real or theoretical system, in order to help people understand or simulate that system

Debriefing
An ethical practice where participants are informed of a study's true purpose and findings once the experiment has ended, and information is given about counselling services if necessary

Deception
An ethical practice where participants are not fully informed about the procedures or aims of the experiment before it is conducted because knowing the purpose might influence their behaviour; they must be thoroughly debriefed afterwards

Dependent variable (DV)
The variable that is observed or measured in an experiment; that which is affected by the experimental condition and is used to measure the effect of the independent variable

Descriptive statistics
Statistics used to summarise, organise and describe data obtained from research

Discussion
In a scientific report or poster, the section where the significance of the findings is examined

Double-blind procedure
An experimental procedure in which neither the participants nor experimenter know who has been assigned to the control and the experimental group(s)

Empirical evidence
Information that psychologists gain from direct observation and measurement; another name for data

Ethics
The moral principles and codes of behaviour that apply to all psychologists in their practice and in their research

Ethics committee
A group, consisting of a range of medical and non-medical professionals, that ensures that the welfare of participants involved in a study is considered

ISBN 9780170465038

Experimental group
The group (or groups) in an experiment that is exposed to the experimental condition(s), where the variable being manipulated (independent variable) is present

Experimenter effect
An extraneous variable that occurs when there is an unintentional change in a participant's behaviour, and hence in the results, due to the experimenter's influence

Extraneous variable (EV)
Any variable other than the independent variable that may cause a change in the results of, and therefore may have an unwanted effect on, an experiment

Fieldwork
An investigation of an issue or line of inquiry in its natural environment

Frequency distribution table
A table used to sort raw data; the categories being compared are placed in one column and the frequencies with which they occur are placed in the other column

Frequency polygon
A graphic depiction of scores from a histogram; a type of line graph

Generalisation
When the findings of an experiment are applied to the population of interest for the research

Graph
A visual display of data that enables large amounts of information to be neatly organised and summarised, and shows the relationship between two variables

Histogram
A graph of continuous data; its bars are always touching

Hypothesis
A testable prediction about the relationship between two or more variables

Independent-groups design
An experimental design that involves randomly allocating members of the sample to either the control or the experimental group(s)

Independent variable (IV)
The condition that an experimenter systematically manipulates or changes in order to gauge its effect on another variable (the dependent variable)

Inferential statistics
Statistics that allow you to make inferences and conclusions about data; often used to interpret results

Informed consent
An ethical principle whereby a researcher must obtain written permission from each participant involved in a study, stating that they consent to participating in the study and have been given all necessary information, including their rights

Integrity
An ethical principle that involves the researcher's commitment to the honest conducting and reporting of research

Introduction
In a scientific report or poster, the section where the background of a topic is introduced and the specifics of the research are presented

Justice
An ethical principle that ensures fair treatment for all, from selection through to result collection

Line graph
A graph consisting of any single line that connects points that relate one variable to another

Literature review
The collation and analysis of the findings of others, or secondary data

Longitudinal study
An investigation into a person or group of people over a period of time, where data are taken at intervals

Matched-participants design
An experimental design that involves pairing each participant with another based on a certain characteristic that they share, and then allocating one to the control group and one to the experimental group

Mean
A measure of central tendency; the number found when all of the scores in a data set are added together and then divided by the total number of pieces of data

Measures of central tendency
Calculations that show how typical scores, or a majority of scores, fall in a data set

Median
A measure of central tendency; the middle number in a data set

Method
The step-by-step approach taken to test a hypothesis

Methodology
The type of investigation chosen to answer a research question; e.g. case study, controlled experiment, literature review etc.

Methodology and method
In a scientific report or poster, the section where the details of methodology, participants, materials and the procedure are stated

Mixed methods research
A study in which both qualitative and quantitative methods are used in data collection or data analysis

Mode
A measure of central tendency; the most commonly occurring number in a data set

Non-maleficence
In the medical profession, the principle of avoiding causing any kind of harm; in scientific research, the implication that, if potential harm/discomfort is involved, the harm/discomfort does not outweigh the potential benefit(s) that could eventuate

Non-standardised procedure
An extraneous variable that occurs when there is a lack of consistency in the procedures used to undertake a test each time the test is administered

Objective data
Data that can be observed and measured

Observational study
A data-collection technique that involves an individual watching a group of people in a natural environment and recording observations about their behaviour

Observer bias
Where an observer in a research study sees what they want or expect to see; may result in an unfair representation of the displayed behaviour

Order effect
An extraneous variable that occurs in a repeated-measures design in which a change in results is due to the sequence in which two tasks are completed

Outlier
A piece of data that lies outside a typical range of scores and can skew the representation of that data set

Participants
The people who take part in a scientific study

Percentage
A mathematical calculation that demonstrates the proportion of a sample that displays a particular behaviour

Personal errors
Errors in research made by the researcher

Placebo
A fake drug or treatment that is used in an experiment so that neither group knows who is being exposed to the experimental condition

Placebo effect
A change in a participant's behaviour due to their expectations regarding the treatment they are receiving

Population
The entire group of people belonging to a particular category that is of research interest

Precision
A gauge of correctness where an assessment tool obtains results consistent with each other

Primary data
Data obtained through new research such as fieldwork, observation or experimentation

Product, process or system development
The act of designing or evaluating a new product, process or system to meet a human need

Qualitative data
Data that describe changes in the quality of behaviour; often expressed in words

Quantitative data
Data, collected through systematic and controlled procedures, that are usually presented in numerical or categorical form

Random allocation
A participant allocation technique that ensures that every member of the sample has an equal chance of being assigned to either the control group or the experimental group in an experiment

Random errors
Errors in research that account for unpredictable variations in results

Random sampling
A sampling technique used in selecting participants for a study, which ensures every member of a population has an equal chance of being selected

Range
A measure of variability that is calculated by subtracting the lowest score in a data set from the highest score

Raw data
The actual data collected from undertaking research

References
In a scientific report or poster, the section that details any materials or publications cited during the writing of the report or poster

Reliability
The extent to which an assessment tool can produce results consistently, each time it is used

Repeatability
The degree to which as assessment tool obtains similar/consistent results when it is conducted again

Repeated-measures design
An experimental design that uses only one group of participants and exposes that group to both the control and experimental conditions

Reproducibility
When the same results occur under changed or different conditions

Respect
An ethical principle shown through consideration for an individual's welfare, but also through the appreciation of their own uniqueness, autonomy and freedom of expression

Results
In a scientific report or poster, the section that features a visual representation of the data that has been collated, as well as a brief written description of the data, but where no analysis is made

Sample
A group of participants selected from, and representative of, a population of research interest

Secondary data
Data sourced through other people's research, obtained from journals or academic articles

Self-report
A data-collection technique where individuals comment on their own thoughts, emotions and beliefs

Simulation
The process whereby a model is used to study the behaviour of a real or theoretical system

Single-blind procedure
An experimental procedure in which the participants do not know whether they have been assigned to the control or the experimental group(s), but the experimenter does know

Standard deviation
A measure of variability that describes how each individual piece of data differs from the mean

Statistical significance
A measure used to determine the likelihood that a set of results occurred due to chance

Stratified sampling
A sampling technique used in selecting participants for a study, which involves breaking the population into strata, or groups, based on characteristics they share, and randomly selecting participants from each stratum in the same proportions that they appear in the population

ISBN 9780170465038

Subjective data
Data based on an opinion; collected through observations of behaviour, or through participants' self-reports

Systematic errors
Errors in research that occur in a consistent manner in relation to the true data value

Title
In a scientific report or poster, a description of the variables that are being manipulated and measured in an investigation

Uncertainty
A lack of knowledge of, or confidence in, the data being measured

Validity
The extent to which an assessment tool actually measures what it is designed to measure

Variability
A mathematical calculation that describes how a set of scores in a data set is spread

Variable
Any condition that can change

Voluntary participation
An ethical principle that ensures that a participant is willing to participate in, or be part of, an experiment

Withdrawal rights
An ethical principle that refers to the right of a participant to leave a study at any time without pressure or negative consequences

Within subjects (design)
An experimental research design in which the same participants are used in both the control and experimental conditions

ANSWERS

1.1

1 *Example answers may include:*

Does working a part-time job while studying increase stress levels?

Did COVID-19 lockdowns increase mental illness diagnosis?

2 An investigation into the impact of gender on aggression

3 a Literature review

b Experiment or case study

c Fieldwork

1.2

Sample answers follow.

Step 1: Identify the research problem

Step 2: Formulate a hypothesis

It is hypothesised that adolescent males who listen to music while studying will score higher on a test of retention of the material studied than adolescent males who do not listen to music while studying.

Step 3: Design the method

Participants will be separated into two groups – one group will study while listening to music; one group will study without music. Participants will be tested on the material studied.

Step 4: Collect the data

Data will be collected by marking participants' tests on the material studied.

Step 5: Analyse the data

The raw data show that test results are higher for participants who listened to music while studying compared to test results for participants who did not listen to music while studying.

Step 6: Interpret the results

The findings suggest that, in adolescent males, listening to music while studying enhances retention of material being studied.

Step 7: Report the findings

1.3

1 a IV: The time the food is given (before or after completing the maze)

DV: The time it takes the rats in each group to find the food/complete the maze

b IV: The flavour of muffin made (either blueberry or chocolate)

DV: The number of compliments received

c IV: The language-learning technique used (hearing and repeating or reading and writing)

DV: Participants' average scores on a language test

2 *Sample answers follow.*

a Experimental design: Select a sample of child participants who do not have any known degenerative eye disease. Test their visual functioning. Split the sample into two equal groups. Have each group watch TV for 1 hour per day for 1 month, but have one group watch TV from a distance of 1 metre and have the other group watch TV from a distance of 3 metres. Retest each participants' visual functioning. Compare the test results.

IV: The distance from which the TV is watched (1 metre or 3 metres)

DV: The difference in visual functioning score between the first and second tests

b Experimental design: Select a sample of child participants and cut a strand of hair from the top of each of their heads, as close to the scalp as possible. Take two measurements: the length of the hair when pulled straight, and the length of the hair when left to fall naturally. Work out the difference between these two measurements for each participant. Split the sample into two equal groups. Have participants in each group eat two pieces of plain bread each day for one month, but have one group eat the bread in its entirety, including crusts, and one group eat the non-crust parts of the bread only. After the month, cut another strand of hair from the same area on the top of each child's head and repeat the measurements, again finding the difference between lengths when the hair is pulled straight and let to naturally fall. Compare the difference in lengths prior to the test with the difference in lengths after the test.

IV: Whether or not crusts were eaten

DV: Average difference in maximum hair length and naturally falling hair length between original and final tests

1.4

1 IV: Undertaking (or not) 10 minutes of meditation before going to bed

DV: Participants' average sleep rating scores

Hypothesis: It is hypothesised that 40-year-old males who meditate for 10 minutes before going to bed will have higher average quality of sleep on a self-reported score than those who do not meditate before bed.

2 IV: The way the class notes are presented (written on board or presented as PowerPoint)

DV: The percentage difference in student examination scores between mid-year and end-of-year exams

ISBN 9780170465038

Hypothesis: It is hypothesised that Year 12 students who are taught using written notes on the board will have lower examination scores than those taught using PowerPoint presentations that include visual cues.

3 IV: The age of participants (20–30 or 50–60)

DV: Participants' average scores on memory-related tests

Hypothesis: It is hypothesised that Victorians aged 20–30 will have higher average scores on memory-related tests than Victorians aged 50–60.

1.5

1 Population: The entire group of people belonging to a particular category that is of research interest

Sample: A group of participants selected from, and representative of, a population of research interest

Random allocation: An allocation technique that ensures that every member of the sample has an equal chance of being assigned to either the control group or the experimental group in an experiment

Control group: The group exposed to the control condition; that is, where the variable under investigation (IV) is absent

Experimental group: The group (or groups) exposed to the experimental condition(s); that is, where the variable being manipulated (IV) is present

2 The control group is used as a basis of comparison to see whether the experimental condition has caused a change in the DV. Without the control group, the researcher would not know if there has been a change, nor what has caused it.

3 *Sample answers follow.*

Control group: Eat burgers in the takeaway franchise with no music present

Experimental group 1: Listen to pop music while eating burgers in the takeaway franchise

Experimental group 2: Listen to classical music while eating burgers in the takeaway franchise

Experimental group 3: Listen to heavy metal music while eating burgers in the takeaway franchise

2.1

1 *Sample answers follow.*

a Participants' existing skills may have differed across the groups – that is, participants in the Friday-night group may have been better tennis players than those in the Saturday-night group, or vice versa.

It may have been windier (or there may have been other impactful weather conditions) on one night than the other, thus influencing serving accuracy.

b There may not be an equal number of males and females in each group, and gender can affect the body's ability to process alcohol.

There may be fewer driving errors due to participants concentrating because they are being watched, or participants may modify their driving behaviour due to using a simulator.

c The mood in each of Ms Luk-Tung's classrooms may differ due to time of day, weather etc., and this mood may affect the way in which students respond to the survey.

As the responses will not be anonymous, students may give positive feedback regardless of their true feelings, for fear of getting in trouble.

2

Extraneous variable
• May influence the dependent variable

Confounding variable
• Known to cause a change in the dependent variable, so its effects may be confused with those of the independent variable

• An unwanted variable
• Impacts on the ability to draw accurate conclusions and generalisations

2.2

1 *Sample answers follow.*

Ask your friends and family.
Ask people at the local shopping centre.
Ask your classmates.

2 *Sample answers follow.*

Gather the names of everyone in Year 12, place them in a hat or box, and draw out 20 at random.

Enter the Year 12 students' ID numbers into a database and sort the database at random, then choose the first 20.

Have each person in Year 12 stand in a long line and select every 3rd person.

3 *Sample answer:* She may split the workplace into four strata: women under 40, women over 40, men under 40 and men over 40. She could then take a random sample from each of these groups in the same proportions that they appear in the company.

4

	Convenience sampling	Random sampling	Stratified sampling
How are participants sampled?	Participants are selected for the sample based on the ease of access and selection.	Participants are selected using a random method so that every member of the population has an equal chance of being selected for the sample.	Members of the population are broken into strata based on particular characteristics. A proportionate number of members in each group are randomly selected for the sample.
Advantages of this method	The sample is very easy to obtain.	It is time- and cost-efficient to select a random sample.	The sample is representative of the population.
Disadvantages of this method	The sample is likely to be biased.	The sample may not actually be representative of the population.	It takes a lot of resources (time and money) to select a sample.

2.3

1

	Explanation	Advantage(s)	Disadvantage(s)
Independent-groups design	Every member of the population has an equal chance of being selected for either the experimental or control group.	Inexpensive, quick and easy to run Can sample large numbers of participants No pre-testing or order effect	May not be a representative sample of the population because no differences between groups have been controlled; i.e. one group may be naturally more intelligent etc.
Matched-participants design	After pre-testing, those with similar characteristics are paired together and one is assigned to the experimental group and the other to the control group.	Many extraneous variables due to participant characteristics are eliminated, as both groups are paired on certain characteristics.	It is time-consuming and costly, as a pre-test must be conducted to match the participants. If one participant is lost from the study, the matching pair must be removed.
Repeated-measures design	The same group of participants is subjected to both the control and experimental conditions.	Extraneous variables related to participant characteristics are eliminated, as the same group of participants is used. Smaller samples can be used.	Order effects can impact on the results.

2 a Repeated-measures design; all participants are exposed to both the control and experimental conditions

   b *Sample answers follow.*

   Interference from the first word list when learning the second word list

   Boredom in learning the second word list due to repetition of the task

   Improved ability in learning second word list due to practice

   c You could expose half of the participants to the control condition (no energy drink) and then the experimental condition (energy drink), and the other half to the experimental condition and then the control condition.

ISBN 9780170465038

2.4

1 a–iii; b–i; c–iv; d–v; e–ii

2

	Similarities	Differences
Placebo and placebo effect	They are both concerned with the influence of participant expectations in research.	The placebo eliminates the impact of the placebo effect on results.
Single-blind procedure and double-blind procedure	In both procedures, participants are unaware whether they are in the control or the experimental group(s).

They both help to reduce extraneous variables in a study (such as the placebo effect). | In a single-blind procedure, the experimenter is aware of which participants are in the control and experimental groups; in a double-blind procedure, they are not aware.

The double-blind procedure eliminates the experimenter effect. The single-blind procedure does not. |

3.1

1 a Correlational study
  b *Sample answer:* It allows for the exploration of a relationship between variables.
  c *Sample answer:* It cannot be used to show causation.
2 a Observational study
  b *Sample answer:* It eliminates the extraneous variable of artificiality (free from participant influence).
  c *Sample answer:* It is subject to observer bias; there are no explanations behind why the behaviour occurs.
3 a Product development
  b *Sample answer:* New technology can help meet human needs.
  c *Sample answer:* It can be difficult to apply broadly due to humans' unique brains.

3.2

1 a Secondary data
  b *Sample answers:* It is less time- and cost-prohibitive than collecting primary data; can be used to make predictions (data modelling)
  c The researchers are warning us about the rates of global warming if greenhouse gas emissions are not reduced, with the purpose of bringing about behaviour change in the present to mitigate these effects.

2

Type of data	Objective/ Subjective	Qualitative/ Quantitative
Hair colour	Objective	Quantitative
Number of hours spent watching TV per day	Objective	Quantitative
What people see in an inkblot test	Subjective	Qualitative
Descriptions of why students like doing homework	Subjective	Qualitative
Average height of students in a Year 10 class	Objective	Quantitative

2 *Sample answers follow.*
  a Verbally ask a sample of students their opinions on why they are 'for' or 'against' wearing a school uniform; have students fill in a survey of open-ended questions about the school uniform.
  b It is difficult to summarise and collate the data so that they can be described and/or compared to other data.
  c Ask students to simply state whether they are 'for' or 'against' wearing a school uniform – do not ask them for reasons.
  d It is still based on students' opinions and therefore cannot be observed or measured accurately.
  e Over a period of time, she could count the number of students who wear the uniform and the number of students who do not wear the uniform, to objectively assess student preferences.

3.3

1 a 7.4
  b 6.5
  c 7
  d The median, because it removes the influence that the top scorer and bottom scorer have on the results and gives a better picture of an average player on the team
  e 28 per cent
  f No, descriptive statistics do not allow conclusions to be drawn. There are many other factors aside from skill that may have contributed to the scoring ability of all team members.
2 a 23
  b 7
  c Sarah's class has high variability, whereas Maeve's class has low variability.
  d No, descriptive statistics do not allow conclusions to be drawn. There are many factors that may have contributed to the performance of each class.

3.4

1 a

Class interval (average no. of goals)	Frequency
0–1	10
2–3	10
4–5	4
6–7	2
8–9	1
10–11	1

b The data are discrete because a player must score a whole number of goals, and the scores are not on a continuum.

c

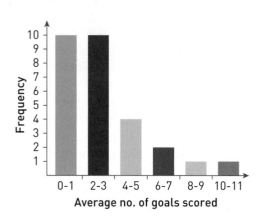

4.1

1 b, d, f, c, e, a

2 a There is a personal error in the data set for the second half of the year: an incorrect/impossible score of 103%. There is an outlier in the scores for the first half of the year: Chase scored 68%, which is significantly lower than his usual score, and is skewing the results. This could either be an error, or it could be that it is a true score but the test was more difficult than usual.

b Chase's average test score in the second half of the year was higher than his score for the first half of the year, which on the surface implies that two hours of study per week is better than one hour. However, the effect of the errors would need to be accounted for, and extraneous variables considered, before any kind of conclusion could be made.

4.2

There are many possible answers. A suggested answer follows.

```
                        A
                        U
                        T
                        H              A
                R       E              P
                E       N              P
V   A   L   I   D   I   T   Y
    C   A               C       O
    T   T                       O
    U   I                       L
    A   O                       S
    L   N
    L   S
    Y   H
        I
        P
        S
```

4.3

1 a Participants were taken from only one country. No cause-and-effect relationship was found. The study did not control variables such as exercise.

b Participants were taken from one city (Manchester, UK). The study did not control variables such as lifestyle factors that could have had an impact on self-reported anxiety scores. The study included only women who were experiencing pregnancy for the first time; anxiety and stress levels of women who already had children may have differed.

5.1

1 *Sample answers follow.*

a Would parents be informed of the long-term impacts that may occur to the child's wellbeing? What support would be available to parents and infants during and after the experiment? How will you ensure that there is no coercion for parental involvement?

b What are the side-effects (type and frequency) of the new medication? What are the long-term impacts of the medication? If the new medication reduces depression and enhances focus and concentration, how will the disadvantages experienced by the placebo group be addressed?

2 Responses will vary.

5.2

1 The lecturer will be using her own students, and therefore has power over them. She has not made any effort to ensure voluntary participation.

Pain is being administered, which causes harm to participants and may create issues with deception in research if not disclosed at the beginning.

2 *Sample answers follow.*

   a Advertise for participants so that the students have to apply of their own accord and free will.

   b After volunteers have expressed their interest, hand out a letter that explains the procedures, participants' rights and the risks involved to the participants. They must sign this before taking part in the study.

   c Be upfront and honest about the procedure wherever possible.

   d Allow participants to cease involvement at any time and withdraw their results at the conclusion of the research.

   e At the conclusion of the experiment, explain all the research and findings to all of the participants and check their wellbeing.

   f Ensure that all items of personal information are kept confidential.

3 *Responses will vary, but could include:* Watson and Rayner's Little Albert experiment, Harlow's rhesus monkey experiments, Seligman's learned helplessness experiments. (Zimbardo's prison study and Milgram's electric shock treatments are already mentioned in this text.)

6.1

1 a An investigation to test the effect of serotonin medication on depression

   b An investigation to test the effect of preparation techniques on confidence when public speaking

2 *Sample answer:* Across the course of a lifetime, an individual experiences stress in all its forms. Stress that is useful to you is known as eustress, while stress that hampers your wellbeing is known as distress. When an individual experiences stress, defined as a psychological and physiological state of tension, it can have health impacts. Research shows us that the psychological experience of stress can result in physiological symptoms. Mild symptoms may include developing a cold or flu, and more extreme symptoms may include developing heart disease. A stressor is any event that can cause stress, and one well-known stressor on young people is academic examinations.

3 1 Theories and definitions; 2 Past research; 3 Aim; 4 Hypothesis; 5 Independent and dependent variables

6.2

Sample answers follow.

- There is no actual methodology mentioned.
- Do not personalise by using words such as 'I'.
- Attachments do not go in the References section; they go in the Appendix/Appendices.
- Individual participant responses are 'raw data', not 'rare data'.
- The 'Participants' section should appear before 'Materials'.
- The 'Participants' section should list the gender of participants, the gender distribution in each group and ages.
- The procedure should be dot-pointed.
- The 'Procedure' should explain what one group did in the experiment while the other group was having acting classes. We may assume from the Results and the Method that the non-acting-class group learnt how to write speeches, but this must be explained and outlined in the 'Procedure'.
- It is incorrect to state that a research report was written as part of the procedure.
- The report writer has presented raw data in the Results; a graph or table of descriptive statistics should be shown instead of raw data.
- There is no statement of what the results show.
- The confidence ratings in the raw data given for each group are identical. It is unlikely that this would have been the case, so one of the sets of data must be incorrect.

6.3

1 *Sample answers follow.*

Extraneous variable: Differences in confidence ratings before any condition was undertaken – some participants may naturally have been more comfortable in public speaking than others.

Future improvements: Conduct a repeated-measures design so that there are no participant differences.

Extraneous variable: The grading of the oral examination could be subjective or influenced by different markers' preferences.

Future improvements: Use a single marker for all presentation assessments, and have strict criteria.

2 a Geoffs, J. (2018). *Brain Surgery for Beginners*. Melbourne: Ebony House.

   b Peters, T. (2021). *Measuring your Emotional Intelligence*. Canberra: Construction Place.

3 • The content of the topic

   • Instructions on how to use memory aids

   • The raw data

   • Permission forms from participants

6.4

There are no answers provided for this task to provide flexibility for use as an assessment or as robust preparation for a future task. All requirements for this task are modelled on the current assessment guidelines for the completion of the scientific poster. Further details can be found at https://www.vcaa.vic.edu.au/curriculum/vce/vce-study-designs/Psychology/Pages/Index.aspx.

7.1

1 *Sample answers follow.*

a IV: Whether or not the pregnant women were offered shopping vouchers

DV: The percentage of women who stopped smoking

Hypothesis: It is hypothesised that English pregnant women who smoke, and who were offered shopping vouchers to quit smoking, would be more likely to stop smoking than those who were not offered vouchers.

b You could select the participants by stratified sampling. The area in which the participants come from (e.g. city vs country) may influence the results. Breaking the women into two groups, 'city' and 'country' (strata) and then selecting a representative proportion from each group for the sample would help to eliminate extraneous variables such as accessibility to shops etc.

Once selected, a matched participants design would be preferable. Not only could you match participants based on their geographical location as described in the sampling procedure, but you could also do it based on the number of cigarettes smoked per day. That way you would balance the effect of heavy vs light smokers, hence minimising the impact of the extraneous variable of individual participant differences, such as difficulty in giving up the cigarettes, or level of addiction etc.

c As smoking is known to have adverse effects during pregnancy, not offering all of the women the highest possible incentive may be a breach of the experimenter's code of conduct to prevent physical or psychological harm. With this in mind, and regardless of the group to which participants were assigned, it would be important to ensure that, as part of their informed consent, participants are given information about (and hence acknowledge through signing the consent form) the dangers of smoking while pregnant.

d As the sample has been selected only from British women, the results cannot be generalised to Australian women.

2 *Sample answers follow.*

a It was hypothesised that Victorian secondary students who used an electronic device in an end-of-year assessment would complete it faster than those who did not use an electronic device.

It was also hypothesised that Victorian secondary students who used an electronic device in an end-of-year assessment would complete it more accurately than those who did not use an electronic device.

b The advantage of using a matched-participants design in this scenario is that extraneous variables such as differences in the participants' academic abilities can be controlled. That is, members of one group will not perform better than members of the other group due to their abilities alone. A disadvantage of this design, in this scenario, is that it would take time to test the students' academic abilities before matching them.

c Although the extraneous variable of participant differences in academic ability has been controlled, there are other extraneous variables that could affect the results. The quality of teaching throughout the year could lead one school to outperform the other school, regardless of the devices used. The subjects the students took could be another extraneous variable, because some subjects may lend themselves more naturally to an examination using either writing or electronic devices. Hence the students may perform better due to the style of the examination rather than the way they complete it.

d As the research involves school-aged students, informed consent would have to be obtained from their parents. They would need to be informed of the potential advantage or disadvantage in doing the exams in this way, and then sign a consent form for their child to participate.

Another potential ethical consideration to be aware of is confidentiality, as the study involves academic results for students and schools. Releasing this information as the results of a study could have a negative impact on a school's reputation or on the perception of a student's abilities.

3 *Sample answers follow.*

a It is hypothesised that first-year university students who take a note-taking skills course will perform better on mid-year examinations than those who do not do the note-taking course.

The independent variable is the type of course taken (note-taking or university resources). The dependent variable is the percentage score on the first semester examinations.

ISBN 9780170465038

b Two courses were offered in this scenario as a means of eliminating the placebo effect. If only the note-taking skills course was offered, participants in this group may have had a change in their results due to their expectation of being involved in a study. To minimise the impact of this, the control group was also given a course, unrelated to note-taking, to eliminate the placebo effect.

The performance of each group would be best displayed as a mean, so the difference in the two group scores could be easily compared.

c Other extraneous variables still uncontrolled in this investigation were that the resources course may have given students extra skills, other than note-taking, that may have been advantageous to their examination performance, and hence this group may have performed more favourably than they would have without any course at all. The two groups were not controlled for academic ability and hence could have featured stronger academic students in the experimental group, which could account for the improvement in results.

It can be concluded that first-year university students who take a note-taking skills course will perform better in their first semester examinations; however, this result only applies to the university from which the sample was drawn.